EXPLORING OUR UNIVERSE

# THE MOON

## Earth's Natural Satellite

EXPLORING OUR UNIVERSE

# THE MOON

## Earth's Natural Satellite

REVISED EDITION

BY FRANKLYN M. BRANLEY
Illustrated by Helmut K. Wimmer

THOMAS Y. CROWELL COMPANY • NEW YORK

*It is our pleasure to thank the following for their help in preparing this volume:*

Dr. Stanley P. Wyatt, University of Illinois, Urbana, Illinois, for reading the manuscript and making suggestions which improved it greatly.

The National Aeronautics and Space Administration for the illustrations on pages 49 and 54–59.

 *Published simultaneously in Canada by Fitzhenry & Whiteside Limited, Toronto.*
*Manufactured in the United States of America*
*L.C. Card 76–146279*
*ISBN 0–690–55415–X*
*0–690–55416–8 (LB)*
*3 4 5 6 7 8 9 10*

*To Peg*

## by the Author

THE CHRISTMAS SKY

EXPERIMENTS IN SKY WATCHING

EXPERIMENTS IN THE PRINCIPLES OF SPACE TRAVEL

EXPLORING BY SATELLITE

MAN IN SPACE TO THE MOON

THE MYSTERY OF STONEHENGE

SOLAR ENERGY

EXPERIMENTS IN SKY WATCHING

### Exploring Our Universe

THE NINE PLANETS

THE MOON: EARTH'S NATURAL SATELLITE

MARS: PLANET NUMBER FOUR

THE SUN: STAR NUMBER ONE

THE EARTH: PLANET NUMBER THREE

THE MILKY WAY: GALAXY NUMBER ONE

# Contents

EXPLORING OUR UNIVERSE

# THE MOON

## Earth's Natural Satellite

# Moonlight

Most people know very little about the moon. They do not understand its motions, the way it produces tides, nor why we had, until 1959, observed only about one-half of its total surface. Yet, it is safe to say that everyone has observed the light of the moon. Everyone has noticed how the light increases as the moon grows, and how bright it becomes when the full moon is riding high, bathing the earth with moonlight.

Long ago people believed that there was some connection between the moon (luna) and lunacy. They believed one should not expose himself to moonlight, for he might come under the spell of luna; he might become a lunatic. Others argued that there was no connection at all between moonlight and one's sanity. To strengthen their argument, these people said that moonlight cannot affect people any more than sunlight can. After all, they pointed out, moonlight is reflected sunlight, and sunlight surely does not drive one mad. Even

though this is true, there are considerable differences between sunlight and moonlight.

| | Per Cent of Radiation in | |
|---|---|---|
| | *Sunlight* | *Moonlight* |
| Ultraviolet light (short wave) | 10 | 1 |
| Visual light | 46 | 5 |
| Infrared light (long wave) | 44 | 6 |
| Planetary heat (very long wave) | 0 | 88 |

Ultraviolet light is invisible. It is the part of sunlight that causes sunburn. Visual light is the light to which our eyes are sensitive. It includes the colors red, orange, yellow, green, blue, and violet. Infrared light has a longer wavelength than visible light. You might con-

The earth from the moon

sider it as the heat from the sun. Planetary heat refers to that part of heat which reflects from the surface of a planet, a satellite, or some other body. The infrared heat referred to comes from a heat producer, whereas the planetary heat comes from a heat reflector.

Moonlight is certainly different from sunlight, but this is no proof that the moon affects the minds of people, that lunacy and luna are related actually.

It is certainly true, however, that the countryside appears strangely beautiful when there is bright moonlight. Indeed, on a clear, bright night when earth is in moonlight, you may have remarked that "it is bright as day." Of course it was not. As a matter of fact, the brightest full moon gives only a fraction as much light as the sun. Indeed, about half a million full moons would be needed to equal the light of the sun. If the entire night sky were packed with full moons, each shining brightly, earth would receive only one-fifth of the light of the sun. The intensity of moonlight is ¼ meter-candle. A meter-candle is the light produced by a candle that is one meter (39.37 inches) away. Moonlight is one-fourth as bright.

This is entirely understandable, for the moon is a dead, inert, and totally cold body. The moon produces no light or heat of its own. It reflects sunlight. We see the moon by the light that it reflects.

In fact, the moon is even a very poor reflector. Astronomers say it has an albedo of 0.07. The word *albedo* comes from the Latin word *albus*, which means "white." Albedo, or whiteness, is the measure of the reflecting power of an object in relation to the amount of light falling upon it. For example, a body would have an albedo of 0.50, or ½, if it reflected half the light that fell upon it. The moon's albedo (0.07) means that the moon reflects 7 per cent of the sunlight that falls upon it. This is low, for Venus has an albedo as high as 0.73, and the albedo of earth is 0.39.

Ninety-three per cent of the sunlight that falls upon the moon is absorbed by the surface. It is converted to heat, which raises the temperature of the lunar surface. The mountains and crater walls cast deep shadows, which affect the moon's reflecting power. If the lunar surface were smooth, free of these deep shadows, the albedo would be increased to about 10 per cent.

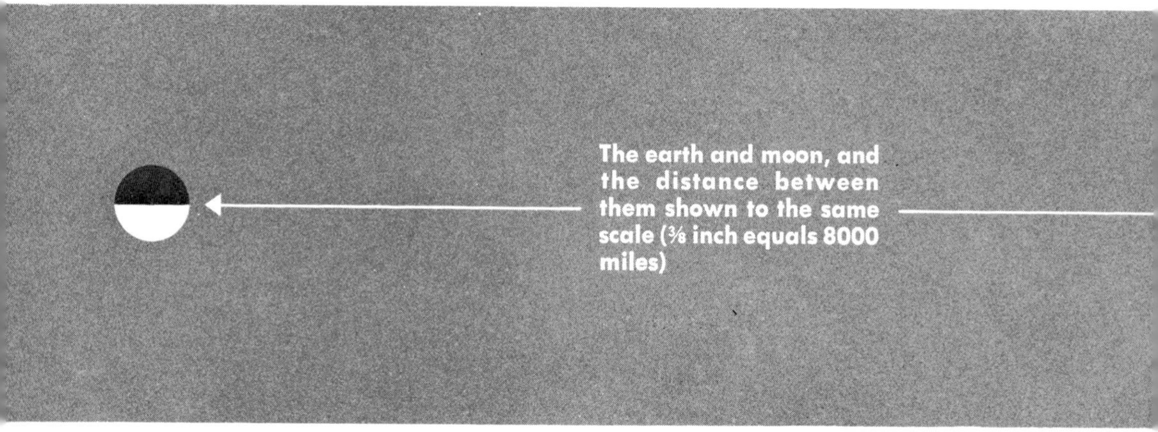

**The earth and moon, and the distance between them shown to the same scale (⅜ inch equals 8000 miles)**

Tests of the reflecting power of various rocks have been made in laboratories. It has been found that dark-colored rocks have an albedo of about 10 per cent. This indicates that the material which composes the lunar surface must be rather dark in color, probably somewhat brownish. However, there are small regions that are much lighter and brighter, and have, therefore, higher albedos. When we say the albedo of the moon is 7 per cent, we refer to the average reflecting power of the whole face of the moon.

Confusion often arises over the manner of referring to the various phases of the moon. Full moon means that we see all of the lighted portion of the moon. (Actually, we are observing one-half of the total moon surface.) Half-moon means we are observing one-half of the lighted portion of the moon. (Actually, we are observing one-fourth of the total moon surface.) The first- and last-quarter phases of the moon are therefore sometimes called half-moons. The half-moon is one-half the area of the full moon.

Since the half-moon is one-half the area of the full moon, we might expect it to be one-half as bright. Actually, it is only one-ninth as bright. One reason is that sunlight which strikes the moon near the terminator, the line separating daylight from darkness, hits at quite a slant, and is therefore not so intense. Another, and perhaps more important, reason, is that the mountains and crater walls cast deep shadows during this phase, thus reducing the surface of bright illumination.

In spite of the fact that bright moonlight is only ¼ meter-candle, some people have eyes good enough to "read a paper by moonlight." If you have ever looked at a magazine by moonlight, you may have noticed that the colors lose their sharpness; they appear as shades of gray.

Colors of inks and dyes depend upon the light that falls upon them. For example, an object appears red in white light because the pigment absorbs some of the wave lengths of the light, and reflects the red. The white light contains all the colors: red, orange, yellow, green, blue, and violet. The pigment subtracts some and reflects others. Suppose the red pigment were illuminated by blue light, what color would you see? You would see no color at all. The red pigment absorbs the blue. Nothing, therefore, would be reflected. The red pigment would appear black.

Make an investigation of colors in moonlight. Gather together magazine pictures that are illustrated in different colors. Take them outside and see if you can identify the colors under bright moonlight. Test other people to see if their results agree with yours. If red appears red, you can conclude that as far as you are concerned, moonlight contains red light. If the red appears black, you can conclude that, as far as your tests indicate, moonlight does not contain red light.

This kind of test is interesting, although from the results one cannot reach any conclusion about the nature of moonlight, for our eyes are not sensitive enough to different wavelengths. The results obtained often do not agree with those obtained by sensitive spectrographs. Spectrographs measure extremely slight variations in wave length. They indicate that moonlight in the visual zone is similar in every respect to light that comes from the sun.

An interesting effect of earth's high albedo often occurs during the crescent phases of the moon. Especially during first crescent, when the moon is low in the western sky in the early evening, we

often see "the old moon in the new moon's arms." The crescent moon appears very bright, and the rest of the lunar disk is also apparent. This happens because sunlight strikes the earth, reflects from earth's surface, and falls upon the moon. It is called earthshine. The way it appears is shown here.

Sometimes the size of the moon seems to change as it rises in the sky. Perhaps you have noticed that the full moon seems to be very large at moonrise. Then, as the moon rides higher in the sky it appears to grow smaller. Some people have tried to explain this apparent change by saying that moonlight is refracted, or bent, by earth's atmosphere. The atmosphere is dense near earth, they say, and at moonrise the moonlight must come through a larger expanse of atmosphere than later in the night. However, this explanation would

not account for the apparent variation in size. Rather, the condition appears to be largely an illusion, for photographs show that the dimensions of the moon at the horizon are no larger than the dimensions of the moon when it is overhead.

Photographs taken of the full moon at various times throughout the night show that the moon's size does not change. Indeed, you can yourself prove quite simply that the moon's size does not change. Hold different objects such as marbles and coins at arm's length until you find one that just covers the rising full moon. Later on, hold the same object at arm's length. You will find that it covers the moon to the same extent that it did earlier, no more and no less.

Probably the moon looks large when it rises because nearby objects appear large; a man in the distance seems much smaller than

Full moon at high elevation

when he is nearby. When the moon is rising we may believe it is closer to us than when it is riding high because we relate it to hills and trees that are close by. And since we know that nearby objects appear larger than distant ones, the moon may seem very large when it is near the horizon.

The landscape pictured in each of the illustrations is exactly the same. The moon is the same size in each illustration. But in one it is near the horizon, and in the other it is high above the horizon. The moon that is near the horizon appears larger than the moon high in the sky.

Things are not always what they seem to be. However, if we search hard enough, we can usually find explanations for the conditions we observe.

# Motions of the Moon

The motions of planets, asteroids, and natural and artificial satellites are affected by nearby bodies. For example, the motions of the moon are affected by the sun and earth. Because the sun, moon, and earth move constantly, their effects upon each other are always changing. Sometimes the sun pulls the moon one way, and then another. Sometimes earth pulls the moon one way, and then another. As a result, the motions of the moon are so complicated that the relative positions of the sun, moon, and earth are probably never repeated exactly. Indeed, there are some eight hundred variations in lunar motions, too many for us to consider here. We shall discuss only the major motions, those which we can observe quite easily. One of these is revolution.

In astronomy, we say a body *revolves* when it goes around another body. We say, for example, that earth revolves around the sun; artificial satellites revolve around earth; the moon, our natural satellite, revolves around earth. Different astronomers may give different times for the revolution of the moon around earth. You may find in some books that the moon takes 29½ days to go around earth. Other sources may tell you that the moon requires 27⅓ days to go around earth. Strange as it may seem, both are correct.

Suppose you walk around a merry-go-round, a carousel. If the carousel were not moving, you could easily tell when you had made a revolution. Just pick a certain horse; go all the way around until the horse is opposite you once more. But suppose the carousel were moving while you were revolving; now the problem is more difficult.

The problem of measuring the revolution of the moon is somewhat the same, because while the moon is going around earth, earth is moving through space.

One method of measuring the time required for the moon to revolve around earth produces 27⅓ days as an answer. An observer on earth sights the moon, lining it up with a distant star. He measures the time that passes until the moon lines up exactly with the same star. This time averages about 27⅓ days; more accurately, 27d. 7h. 43m. 11.47s., or 27.32166d. This is called the sidereal month. The word *sidereal* comes from the Latin word *sidus*, meaning "constellation" or "star group." A star can be used as a fixed point in the sky from which to measure motions of other objects because stars move very little in the short period of a lifetime. We say that the stars are fixed, although astronomers know that they really are not. Stars move constantly, but because they are so far away we are not aware of the motion.

Another method of determining the revolution period of the moon is to measure the time that passes between a phase and the next appearance of the same phase. For example, about 29½ days are needed for the moon to pass from new moon through first quarter, full moon, last quarter, and back to new moon. This is because the phases of the moon depend upon the positions of three bodies: sun, moon, earth. For example, new moon occurs when the sun, moon, and earth are in line. While the moon goes around earth, earth goes part way around the sun. The moon takes about two days to catch up; it takes

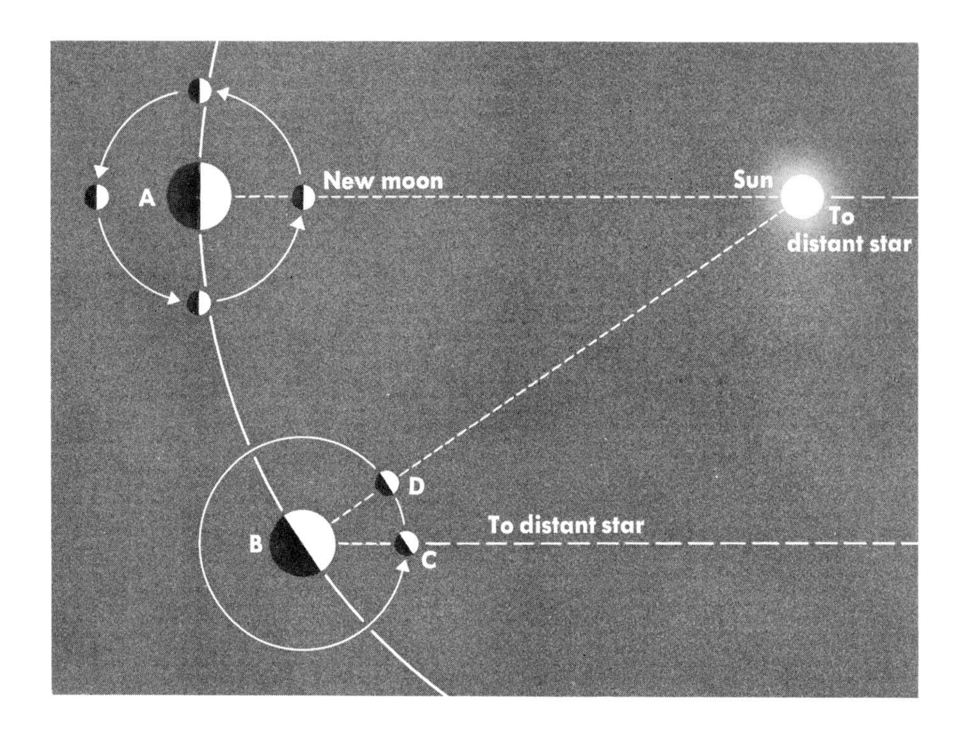


about two days to get into the new-moon position. This is called a synodic month, after the word *synod*, meaning "convention" or "assembly." The average length of a synodic month is 29d. 12h. 44m. 2.78s., or 29.53059d.

In the top of the diagram, earth is at A. The moon lines up with the sun and also with a distant star. The earth moves along its orbit, the solid line. Later, it is at position *B*. During this time, the moon has gone around earth. In 27⅓ days, the moon is at C. Once more it is in line with the same distant star, and a sidereal month has passed. It takes almost two more days for the moon to get to position *D*, that is, to reach the new-moon position. We add this time to the sidereal month, and we have about 29½ days for the synodic, or phase-to-phase month.

Revolution of the moon around earth causes the moon to move from west to east across the sky. The moon goes all the way around earth in 27⅓ days; it completes a circle. There are 360 degrees in a circle; therefore, in one day the moon moves through 360 degrees divided by 27⅓, or about 13 degrees. If you line up the moon carefully with a star, you will find that the next night the moon will be about 13 degrees east of that position. You can approximate such a distance by remembering that the distance between the pointers of the Big Dipper is 5 degrees.

As the moon moves from west to east, its appearance changes. The new crescent moon, no more than a sliver, appears low in the western sky right after sunset. The moon moves eastward, so the next night it appears a bit higher in the west; it sets about one hour later. Also, the crescent is slightly larger. As the days pass, the moon appears farther and farther toward the east. About two weeks after the first crescent is seen, the moon is in the eastern sky at sunset. The moon rises as the sun sets, and now the moon is full.

The apparent changes in the shape of the moon are called the lunar phases. They occur because the moon moves around earth. The moon is a sphere, so one-half of it is always lighted by the sun, unless the moon is in earth's shadow.

During new moon, the lighted half is turned away from earth, and therefore we cannot see any of it. New moon is completely invisible.

An evening or so later, the moon has moved part way around earth, and now we can see a small part of the lighted half. After a week, the moon is one-fourth of the way around earth. We are able to see one-half of the lighted portion of the moon, or one-fourth of the total lunar surface. This is first quarter moon.

After two weeks, the moon is on one side of earth and the sun is on the other side. When the moon is not in earth's shadow, we can see all of the lighted half of the moon. We say the moon is full.

As the moon gets older, it appears to decrease in size until, after another week, it appears once more as a quarter moon. It is last quarter. The moon continues to decrease until only a small crescent can be seen. The next night the moon is new. Once again it is invisible.

When the moon is growing, moving from new to full, it is called a waxing moon. From new to first quarter, it is called a waxing crescent. From first quarter to full, it is called a waxing gibbous moon.

When the moon is decreasing, as when it moves from full to new, it is called a waning moon. From full to last quarter, it is called a waning gibbous moon. From last quarter to new, it is called a waning crescent.

An observer out in space, one who was far above the earth-moon system, would not see the phases, nor would he see the moon moving around earth in the way the motions are usually shown. Rather, he would see both earth and the moon moving around the sun, as shown in the drawing on pages 16 and 17.

The moon is close to earth. It is only about 1/400 of the distance from earth to sun. Also, earth moves very fast in its path around the sun; about 18.5 miles per second. The moon moves much more slowly around earth; about 0.6 of a mile each second. Because of the closeness of the moon, and because of the great distance to the sun, the path of the moon about the sun (as seen by an observer in space) would always be concave to the sun.

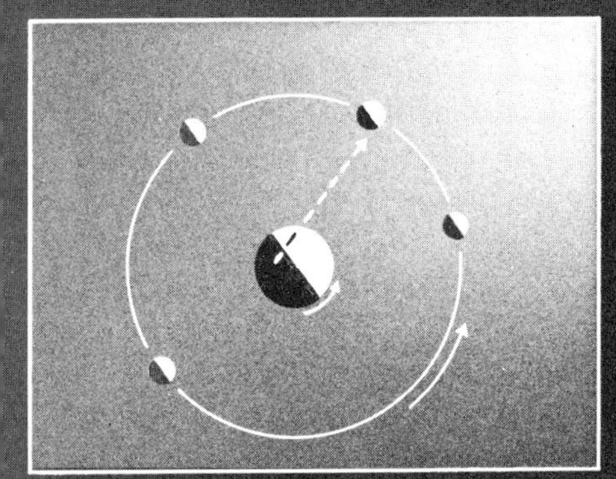

Waxing crescent

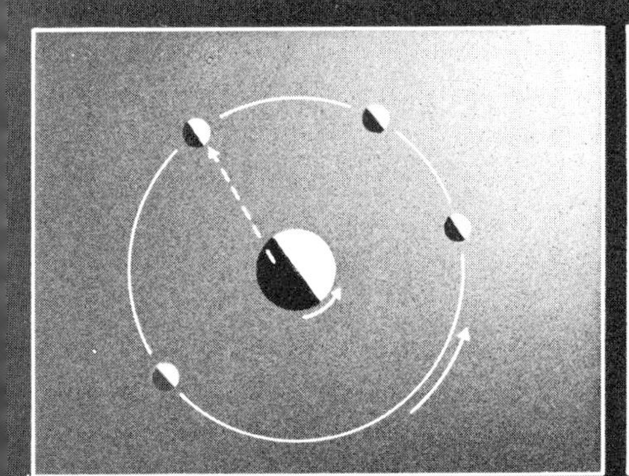

First quarter

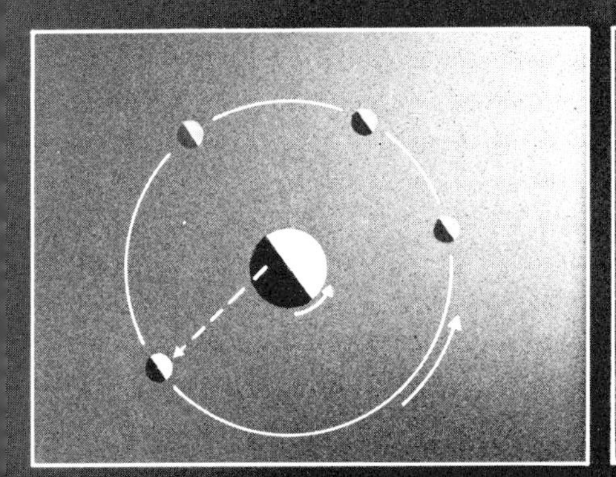

Full moon

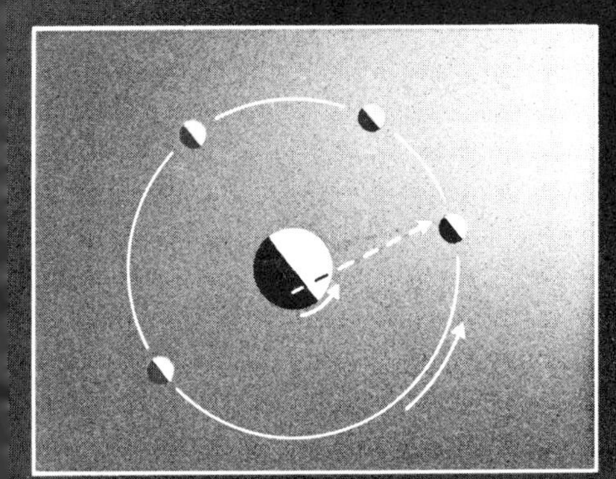

Waning crescent

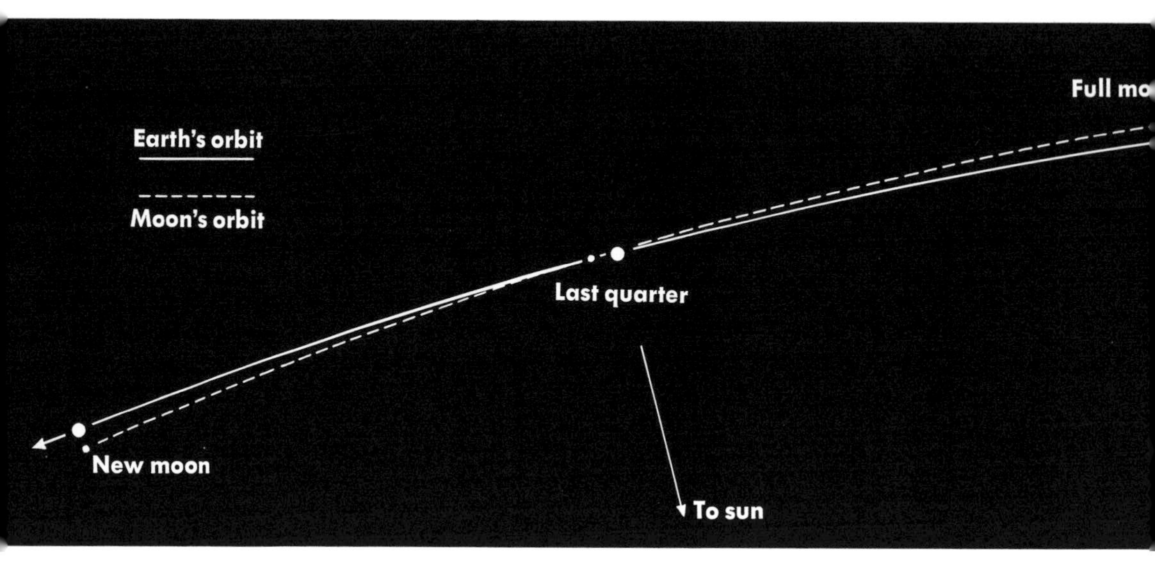


A model that demonstrates this relationship might be made as follows: Connect newspapers with tape to make a sheet of paper a bit over 100 inches square. Using a 100-inch piece of string, draw a quarter circle on the paper, using one corner of the paper as the center of the circle. This represents the orbit of earth about the sun. Draw the path of the moon about the sun, being sure that it never deviates by more than ¼ inch from earth's orbit. When the moon is on the far side of earth's orbit from the sun, a full moon can be seen from earth. When the moon is on the near side, there is a new moon.

*Rotation* is another of the major motions of the moon. In astronomy, rotation means the motion of a body around its axis. For example, earth rotates on its axis. In a similar fashion, the moon rotates on its axis, producing day and night on the moon.

Compared to the speed of rotation of earth, which is about 1000 miles an hour at the equator, the moon rotates very slowly indeed. The moon rotates about 10.35 miles per hour at its equator. The moon makes one complete rotation in a sidereal month. Because the

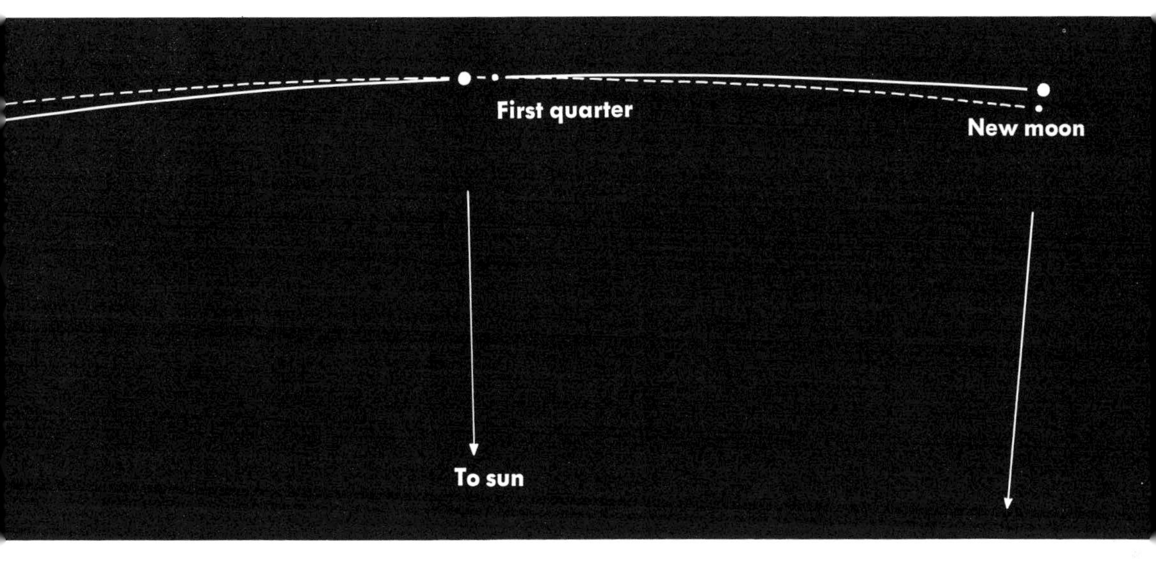


moon rotates and revolves in the same length of time, the same half of the moon is always turned toward earth. Actually, we see a little more than half of the moon.

A simple experiment will show why we always see the same half of the moon. Stand in a room. Face forward, and turn around once completely. As you turned, you faced all four walls of the room. In other words, you faced all four directions, north, west, south, and east; you made one complete rotation.

Now place a chair in front of you; the chair represents earth. Go all the way around the chair—that is, revolve around it—always facing the same wall. You revolved, but you did not rotate. Now go around the chair, facing the chair at all times. Or, go around the chair while facing each of the four walls in succession. This time you revolved once around the chair, and you rotated once. The same half of your body was always toward the chair. Someone sitting on the chair (or on a stool so that he could turn easily if he wanted to) would never see the other half of your body. And so it is with the earth and the moon; from the earth we cannot see the other half of the moon.

In reality, we have seen 59 per cent of the moon. We can see a bit more than half of the moon because of libration. The word "libration" comes from the Latin *libra,* which means "to balance." Galileo, because he was the first person to look at the moon through a telescope, was therefore the first person able to observe the moon with great care. He discovered these slight variations in the viewed portion of the moon.

## Libration in Latitude

Latitude on earth is distance north or south of the equator. On the

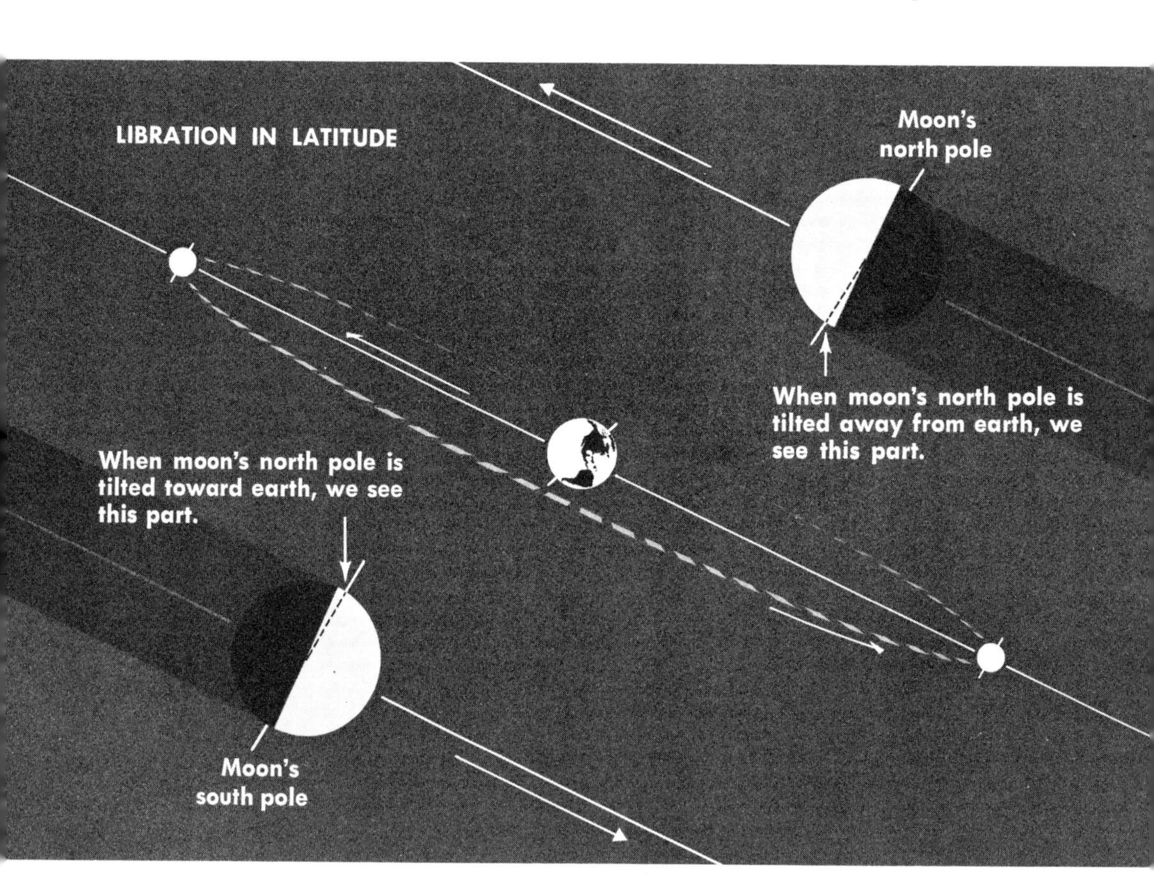

moon, latitude is distance north or south of the equator of the moon. Therefore libration in latitude must refer to variations north or south of the equator.

The axis of the moon is tilted slightly to its path through space. We say that the axis of earth is tilted 23½ degrees from a line perpendicular to earth's path through space. The axis of the moon is tilted 6½ degrees to its path through space. Therefore, when the axis is tilted *toward* us, we can see 6½ degrees beyond the moon's north pole. When the axis is tilted *away* from us, we can see 6½ degrees beyond the moon's south pole. The moon seems to rock from north to south.

## Libration in Longitude

Longitude on earth is distance east and west of the prime meridian, an imaginary line running from the North Pole to the South Pole, and passing through Greenwich, England. Longitude on the moon has a similar meaning. You might call it distance around the moon at right angles to imaginary lines drawn from the North Pole to the South Pole. By agreement, it is measured east and west from the meridian that passes through the eastern edge of the Sea of Rains and through the Sea of Vapors.

As the moon goes around earth, it does not always go at the same speed. The moon's orbit is an ellipse; and, when the moon is near earth, the moon revolves faster than it does when farther away. We shall see why shortly. However, the *rotation* speed of the moon does not change. Because of this difference, the revolution of the moon does not keep pace with rotation. When the moon moves one-quarter around earth, it should rotate one-quarter. But sometimes it rotates a little more than one-quarter, and sometimes a little less than one-quarter. Therefore, we are able to see a little farther around the moon on one side, actually about 6 degrees toward east or west. The moon seems to rock from east to west.

Suppose an arrow were erected on the moon along the line that connects the center of the moon with the center of earth. This is

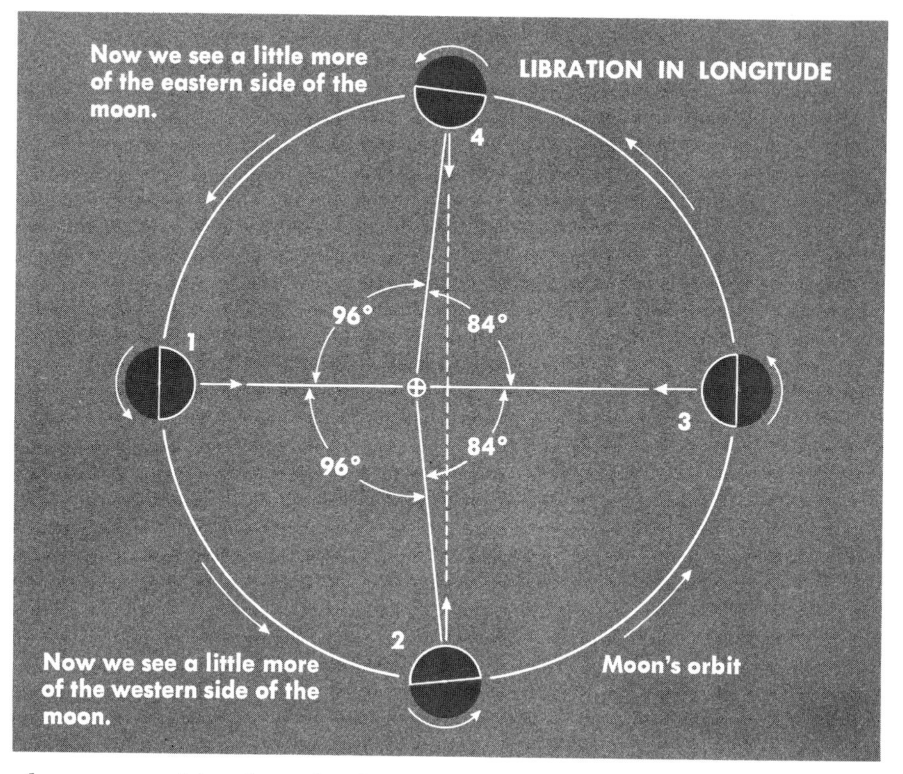


shown at position 1 in the diagram. Viewed from the center of earth, the arrow would be at the exact center of the disk of the moon. The moon is at perigee, the location when it is closest to earth.

After a quarter of a month has passed, the moon has moved to position 2; rotation of the moon will change the direction of the arrow by 90 degrees. But during this same interval, the moon has revolved a little *more* than one-quarter of the way around earth. The upright arrow appears a little east of the center of the moon. Now we can see a little of the western side of the moon; a part which could not be seen when the moon was at position 1.

After half a month, the moon has moved through 180 degrees; it is at the opposite location. Rotation has caused the moon to turn 180 degrees on its axis, and so the arrow once again appears to be at the center of the lunar disk.

After three-fourths of a month, the moon has rotated three-fourths of the way around its axis, or 270 degrees. However, the moon has revolved a little less than three-fourths of the way, only 264 degrees. Now the arrow appears a little toward the west of the center of the moon. We can see a little of the eastern side of the moon, a part we could not see before.

## Diurnal, or Daily, Libration

This variation in visibility is caused by the rotation of earth. When the moon is rising, we see slightly over the top of it. This is actually the western edge of the moon. Similarly, when the moon sets, we can see slightly over its eastern edge. In the diagram, the dotted lines show the half of the moon that is visible when it is overhead. The portion on the earth side of the solid line is the part that we can see at moonrise and at moonset.

The total effect of all these librations is to increase the amount of

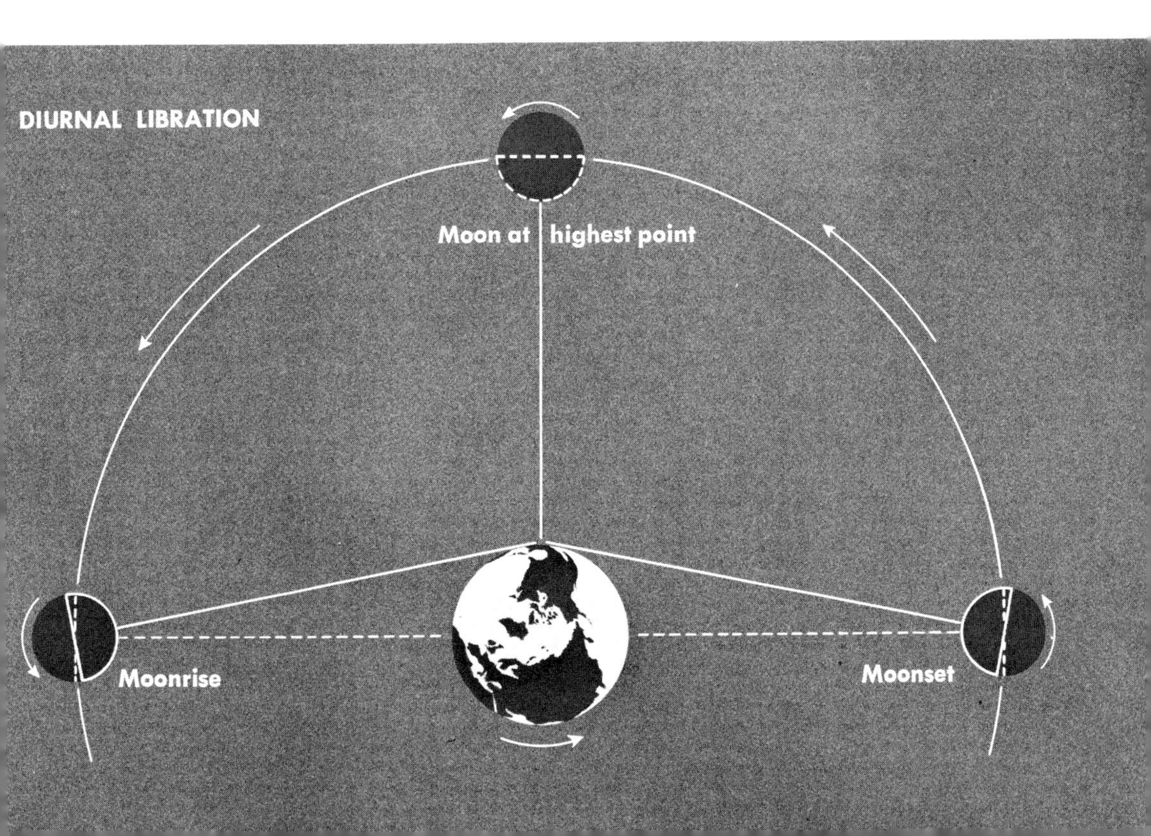

the moon that we can see to about 9 per cent more than half. Forty-one per cent of the moon never disappears from view. Another 41 per cent of the moon is never visible from the earth. Eighteen per cent is alternately visible and invisible.

| | |
|---|---|
| 41% | not seen |
| 41% | always seen (when moon is visible) |
| 18% | alternately seen and not seen |
| 100% | |

But why, you may ask, does the moon sometimes cover less than one-fourth of its orbit in a quarter of a month, while on other occasions it covers more than one-fourth of its orbit in the same length of time? Does the velocity of the moon change; does the moon move faster at some times than it does at other times? Indeed, it does.

The path of the moon around earth is elliptical; sometimes the moon is only 221,463 miles from earth's center, at other times it is 252,710 miles away.

Johannes Kepler, a famous German astronomer who lived in the seventeenth century, figured out some amazing facts about the orbits of the planets and the moon. The planets, he said, and the moon, move in elliptical orbits; and he also said that a line joining earth and the moon would sweep through equal areas in equal periods of time. Let's see what this means. In the diagram, suppose the moon is at A. Five days are needed for the moon to move to *B*. Later on, the moon is at C. Five days are needed for the moon to move to *D*. You can easily see that *AB* is longer than *CD*. Yet the same amount of time is needed for the moon to move from A to *B* as to move from C to *D*; therefore, the moon has to move faster when it goes from A to *B*. When the moon is in the *A-B* region, it is near earth (perigee), therefore it moves faster. When the moon is in the *C-D* region, it is far from earth (apogee), therefore it moves slower. The speed must vary this way, because Kepler proved that a line from earth to moon must sweep through equal areas in equal periods of time; in other words, area 1 must equal area 2.

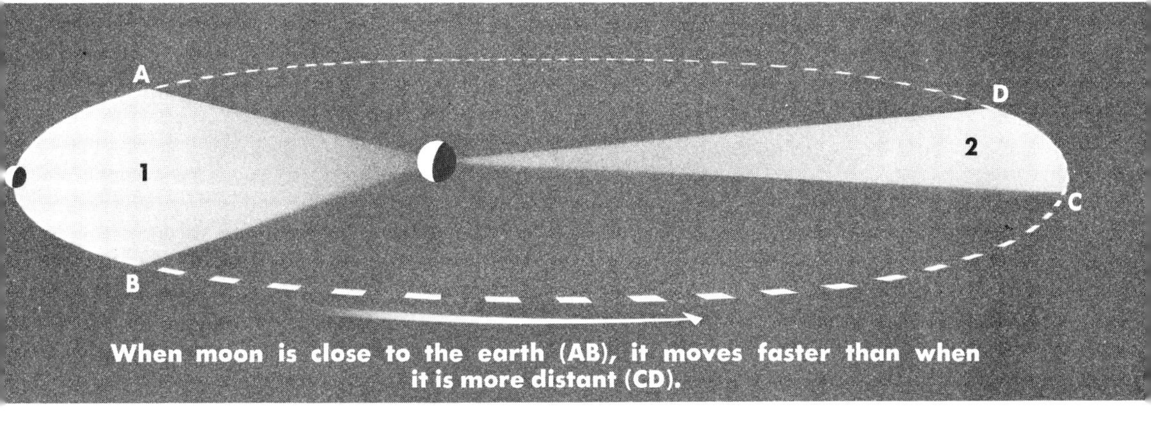

When moon is close to the earth (AB), it moves faster than when it is more distant (CD).

The velocity of all the planets and of all the natural and man-made satellites varies in a similar manner.

## Harvest Moon and Hunter's Moon

The moon appears to rise in the east, move across the sky, and set in the west. This *apparent* motion is caused by the turning of earth. The *actual* motion of the moon around earth is just the opposite, from west to east. During a night and day the moon moves eastward about 13 degrees and earth must therefore rotate a bit longer than 24 hours before the moon rises above the horizon the next night. This increased time averages about 50 minutes. Astronomers say that the retardation in moonrise averages about 50 minutes.

However, the retardation of time varies a great deal. In the fall, moonrise can occur only about 22 minutes later each night, while it may be 80 minutes later in springtime. This is why:

The moon moves across the sky along essentially the same path that the sun travels. The path is called the ecliptic, because eclipses occur along this line. During the new moon phase, the sun and moon are on the same side of earth, and they are at essentially the same location on the ecliptic. During full moon, the sun is on one side of

During winter (right) the full moon is high in the sky.

earth at a certain location on the ecliptic; and the moon is on the other side of earth, at the opposite location on the ecliptic. Below, A, B, C, and D are at the same location on earth but rotating. At A an observer sees the full moon and the planets high, but at B the new moon and the sun appear low. On the other hand, at D the full moon and the planets are low and at C the new moon and sun are high. The diagram above shows the relationship with the ecliptic.

If we extend earth's equator out to the sky, the imaginary line that is formed is the celestial equator. The ecliptic is tilted 23½ degrees to the celestial equator and so the two lines intersect. One intersection is in the constellation Aries, and this is the spring equinox, or the position that the sun occupies on the first day of spring. The other intersection is in the constellation Libra; it is the fall or autumn equinox, or the position that the sun occupies on the first day of fall.

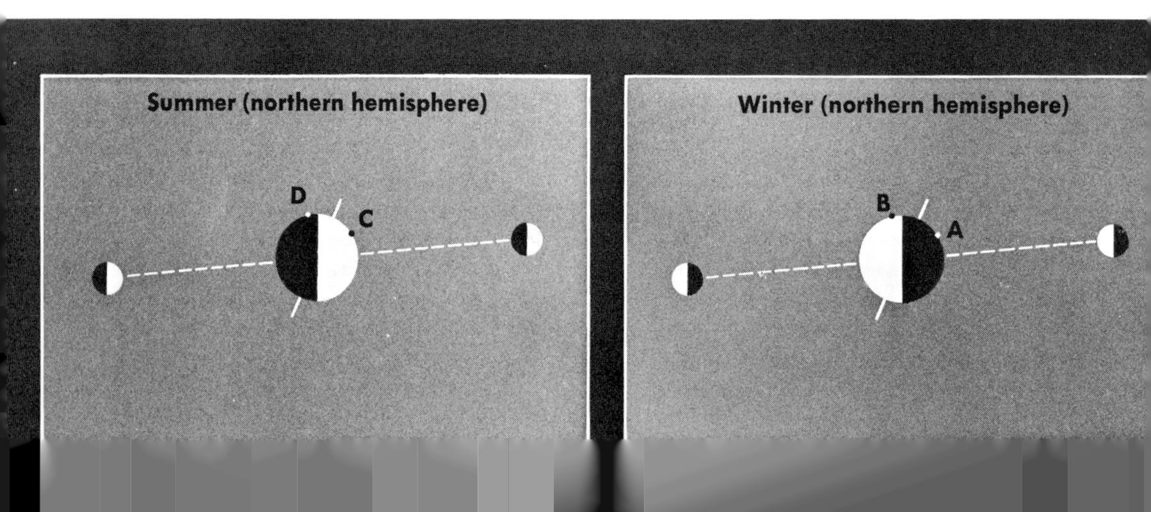

The harvest moon occurs in September, around the beginning of fall when the sun is in Libra and the moon is in Aries. The hunter's moon is the full moon that follows the harvest moon. Each of these moons has a short retardation of moonrise from night to night, and so farmers and hunters of long ago had moonlight shortly after sunset to harvest crops and hunt game late into evening.

During fall, moon's orbit (the ecliptic) makes the least angle with the horizon. This is because the part of the ecliptic that is above the horizon lies south of the celestial equator. Therefore, the daily motion of the moon along the ecliptic has the minimum effect in delaying moonrise.

In spring, just the opposite is true. The part of the moon's orbit (the ecliptic) above the horizon is north of the celestial equator.

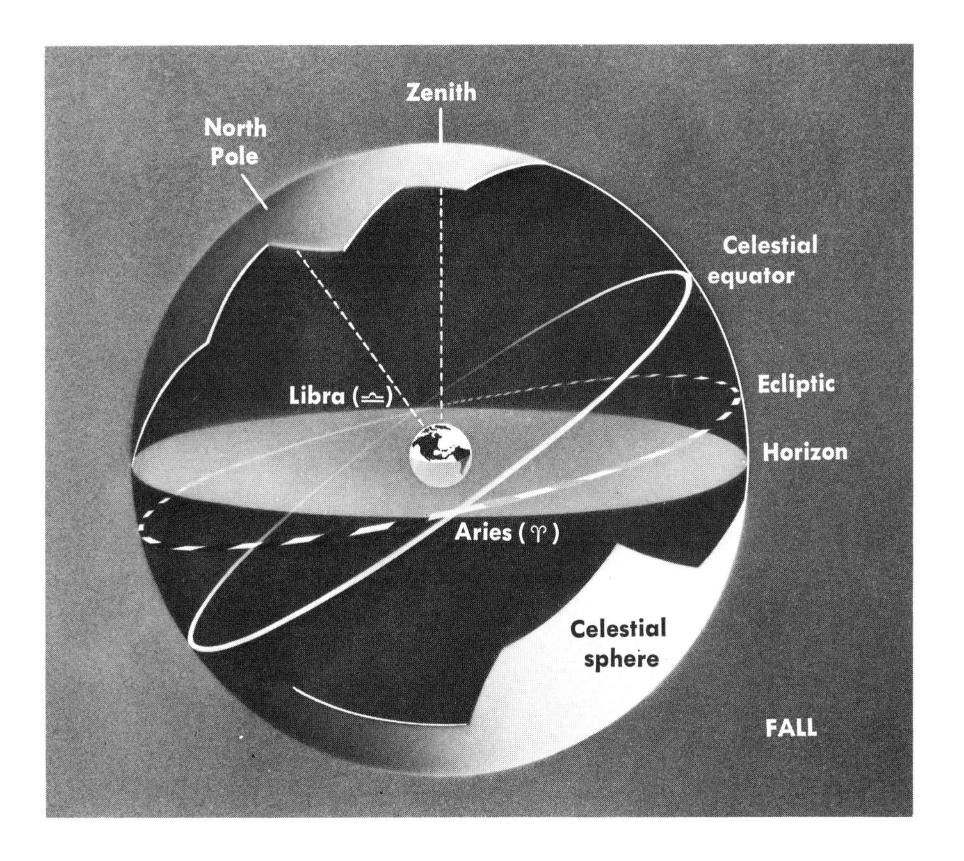

Study the diagram explaining it on the next page, and you will see that the daily motion of the moon along the ecliptic in springtime causes the time of moonrise to be greatly retarded.

In the fall suppose the moon rises at position 1 at 6:00 P.M. During the next 24 hours it moves from 1 to 2 on the ecliptic. By 6:00 P.M. the next night it is at 2, below the horizon. It may take about 22 minutes for the horizon to drop to 2, so that the moon rises at 4 at

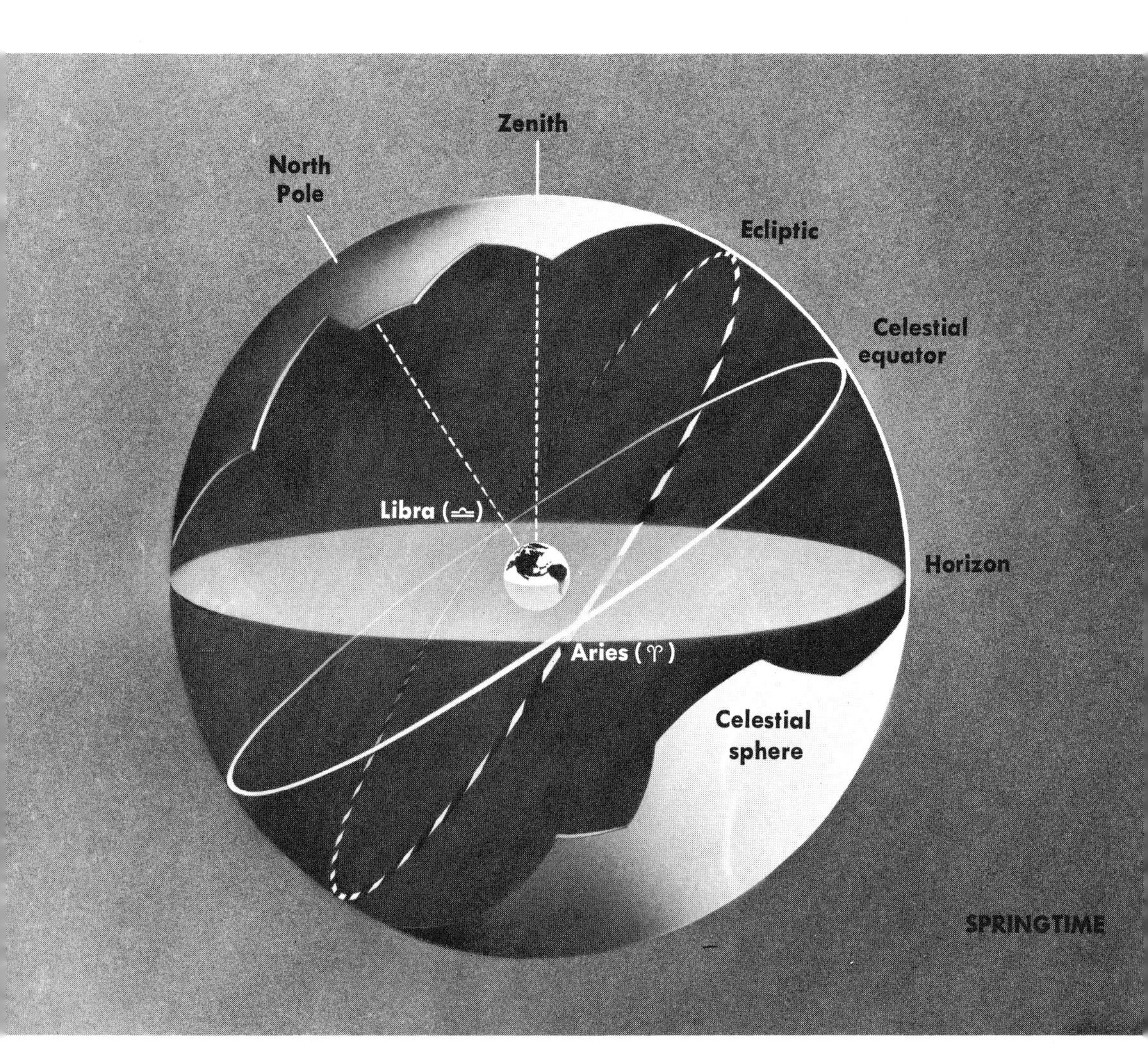

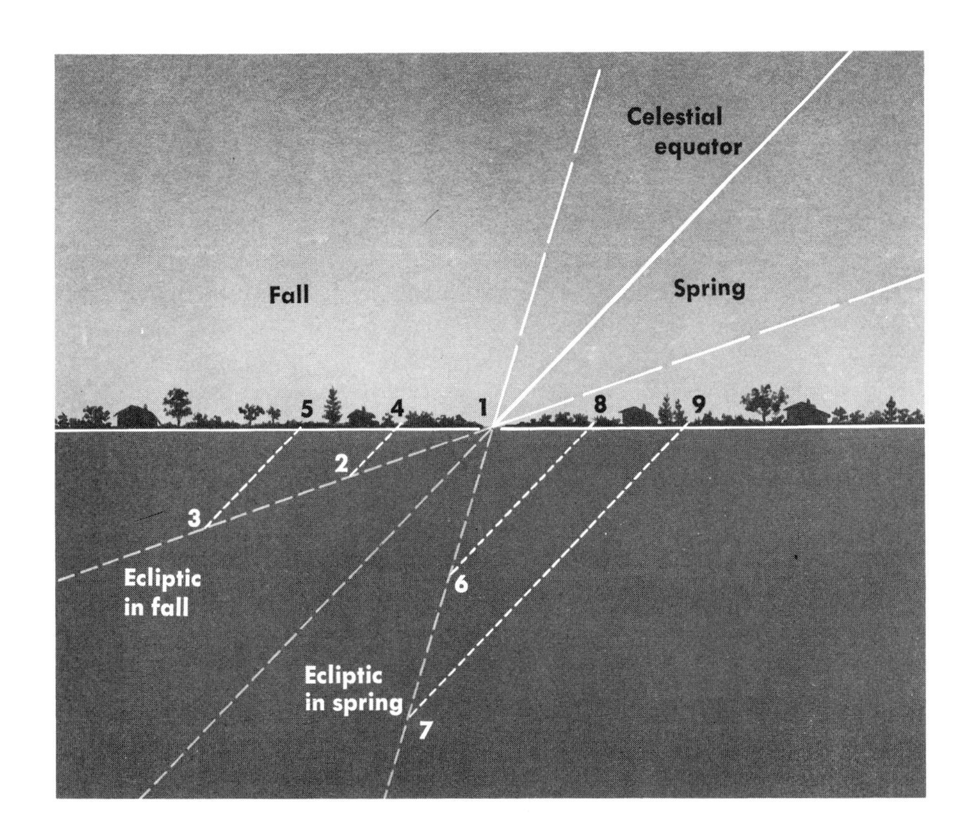


6: 22 P.M. During the next 24 hours this is repeated. The moon is at position 3 by 6:00 P.M. the third day and rises at position 5 at 6:44 P.M.

In spring suppose the moon rises at position 1 at 6:00 P.M. By 6:00 P.M. the next night the moon has moved along the ecliptic to position 6, but it will be 60 minutes before the horizon drops to allow the moon to rise at 8 at 7:00 P.M. The third night it will be at 7 at 6:00 P.M. but will not rise at 9 until 8.00 P.M.

# Why the Moon Stays in Orbit

It is easy indeed to understand how the horses go around a carousel, or how a weight can be swung around on a string. It is easy, because the horses are attached to the platform, and because the weight is attached to the string. It is more difficult to understand how the planets move about the sun, or how the moon moves about earth, for there is nothing visible connecting the bodies.

Every 27⅓ days the moon travels some 1,396,000 miles in an orbit that circles earth. To move that far, the moon must move at a velocity of about 2287 miles per hour. No one knows positively what caused the moon to revolve around earth originally. We do know it is moving now, and we believe it will continue to move. Why does the moon continue to move around earth?

To find an answer, we must turn to Sir Isaac Newton, the famous English scientist who lived from 1642 to 1727. Newton made many important discoveries. For example, he discovered that all bodies exert a gravitational attraction upon all other bodies, and that the attraction increases as the mass increases.

Suppose there were two bodies a thousand miles apart. What would happen if the bodies were twice as far apart? Newton found the answer. If bodies are twice as far apart, the gravitational attraction willl be one-fourth as great: the force varies inversely with the square of the distance.

This is Newton's Law of Universal Gravitation: every particle of matter attracts every other particle with a force proportional to the product of their masses and inversely proportional to the square of the distance between them. The mathematician expresses it this way:

$$F = G\,Mm/r^2$$

where $F$ = gravitational attraction
$G$ = constant of gravitation (that is, the force between two given masses at a given distance)
$M$ = mass of one body
$m$ = mass of another body
$r$ = distance between the two bodies

After the discovery of the Law of Universal Gravitation, scientists made rapid progress in their understanding of problems that had perplexed them for generations. One of these problems was how and why the moon goes around earth.

Suppose we apply Newton's law to the earth-moon system. In diagram 1 (page 30), the moon, $M$, is moving around earth, $E$. Let $M$ represent moon's mass, and $E$ represent earth's mass. From Newton's law we know that there will be a gravitational attraction of $E$ upon $M$. In the diagram we show this as $F$. The direction of this force is along the line $E$-$M$ that connects the two bodies. If $r$ (the distance between the two bodies) increases, the force $F$ will decrease. The force $F$ will never become zero; therefore, the moon ($M$) is always pulled toward earth ($E$). Also, as long as the moon moves in a circular orbit, the distance between earth and moon will remain unchanged. Therefore, the force ($F$) will not change.

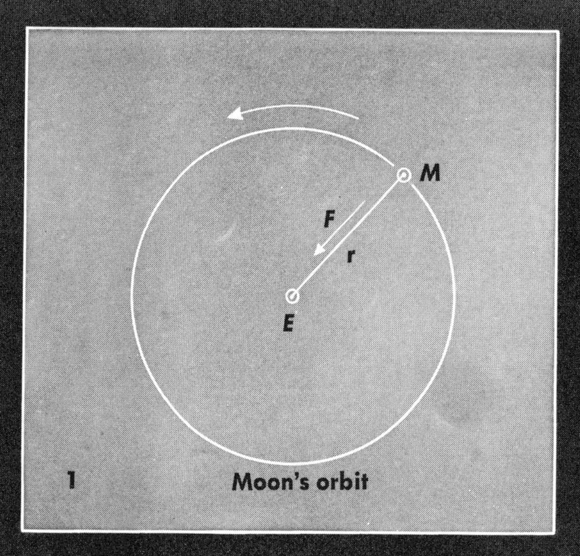


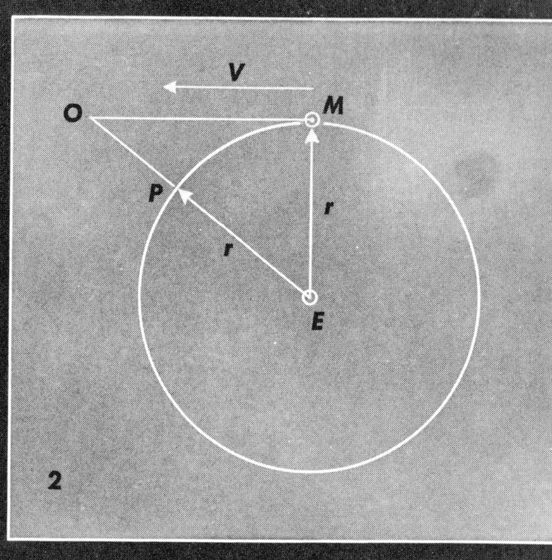


E = mass of earth M = mass of moon

F = gravitational attraction of E upon M

r = distance between earth and moon

OP = distance moon falls toward earth in a given time interval

V = velocity

If earth is always pulling on the moon, why does the moon not fall into earth? Newton answered this question when he stated his Laws of Motion. Newton's laws said that a body at rest remains at rest unless a force is exerted upon it; a body in motion remains in motion and travels in a straight line unless a force is exerted upon it.

This means that if no force (a push or pull, for example) is exerted on a body, the body remains still or moves in a straight line. If several forces are exerted, and if the forces are balanced, the same thing happens. For example, if you were pushing a car one way, and if someone else were pushing it the other way with equal force, the effect would be zero. One force offsets the other. Similarly, suppose a steel ball were rolling along. One magnet pulls the ball to the right, another pulls it to the left. If these magnetic forces are exactly equal, the ball will move in a straight line; the forces offset each other. To cause the ball to move in a curved path, one force must be greater than another—the forces must be unbalanced.

If something moves in a curved path, it means that the forces exerted upon it must be unbalanced because, if the forces were balanced, the object would move in a straight path. The moon moves in a curved path, so the force on it must be unbalanced. The force exerted upon the moon is the gravitational attraction of earth. This force is not balanced by any other force.

Newton gave us another Law of Motion. He said that change of motion is proportional to the force exerted on a body and that it takes place in the direction in which this force is exerted upon it. This simply means that if you push toward the right, a moving object goes to the right. If you push hard, it turns sharply. This is called the Law of Acceleration. When an object is accelerated, it may move faster or slower; it may move in a different direction; or it may do both.

How does this apply to the earth-moon system? We have said that the only force exerted on the moon is the gravitational attraction of earth. This force is toward earth (*E*) ; and so the moon (*M*) must be accelerated toward earth.

But how can the moon be accelerated toward earth when its distance and speed around the earth do not change? (*Note:* We are assuming a circular orbit to simplify the explanation.) It can do this because acceleration here does not change the *velocity*; it changes only the *direction*. Look at diagram 2.

We already know that *M* tries to move in a straight line—*MO*, for example. (Newton said objects in motion move in straight lines.) But a force is exerted on *M*. In the time it takes *M* to move to O, earth pulls the moon from a straight line. The amount that the moon falls toward earth in this time is shown by the line *OP*. The result is that the moon moves on the curve *MP* around earth. If the distance *OP* is just right, the moon moves in a circular orbit around earth.

The unbalanced force of earth's attraction on the moon causes the moon to fall toward earth, to pull away from a straight line. The moon is falling constantly toward earth; the amount of fall is 1/19 inch each second. If it did not fall toward earth, the moon would move along a straight line into space, never to be seen again.

For the same reasons, earth moves in a curved path about the sun. Each second earth falls ⅛ inch toward the sun. It moves in a curve because the gravitational attraction of the sun deflects it from a straight path.

You can explain how man-made satellites stay in orbit by following the same line of reasoning. This is an interesting example of how modern man applies basic ideas that were discovered and explained centuries ago.

# Eclipses

Down through the ages eclipses have intrigued men. Ancient men were ignorant of the causes of an eclipse, so they were frightened by them. The early Chinese, for example, believed that a great dragon devoured the sun. The only way to get the sun back again was to make loud noises to frighten the dragon. If you frightened him, the dragon would disgorge the sun, and once more the day would be bright.

We understand the causes of eclipses, so we are not fearful of them; we can enjoy their beauty. Many of the readers of this book probably will have seen an eclipse. If you have not, do not despair, because an average of four eclipses occurs each year. There always are at least two solar eclipses a year. There may be as many as seven eclipses, five solar and two lunar, or four solar and three lunar.

Even though solar eclipses are more numerous over earth as a whole, we see lunar eclipses more frequently. This is because a lunar eclipse can be seen over considerably more than one-half of earth, while solar eclipses can be seen from only a very limited area, a band a hundred miles wide, more or less, and three or four thousand miles long.

## Lunar Eclipses

Lunar eclipses will occur between 1972 and 1982 on the dates shown in the accompanying table.

| Year | Date | Type |
|---|---|---|
| 1972 | January 30 | Total |
| | July 26 | Partial |
| 1973 | December 10 | Partial |
| 1974 | June 4 | Partial |
| | November 29 | Total |
| 1975 | May 25 | Total |
| | November 18 | Total |
| 1976 | May 13 | Partial |
| 1977 | April 4 | Partial |
| 1978 | March 24 | Total |
| | September 16 | Total |
| 1979 | March 13 | Partial |
| | September 6 | Total |
| 1981 | July 17 | Partial |
| 1982 | January 9 | Total |
| | July 6 | Total |
| | December 30 | Total |

A lunar eclipse occurs when the moon moves into earth's shadow. Such an eclipse can happen only during the full moon phase. But we do not have an eclipse during each full moon. This is why:

Earth goes around the sun in a certain plane. For example, suppose the sun were at the center of a tabletop. Earth goes around the sun on the level of the tabletop. At the same time, the moon goes around earth, but the moon's path is tilted a little more than 5 degrees to the tabletop, the plane of earth's orbit. Although earth always casts a shadow, we are not aware of it because the moon usually passes above or below earth's shadow. If the moon happens to be near the

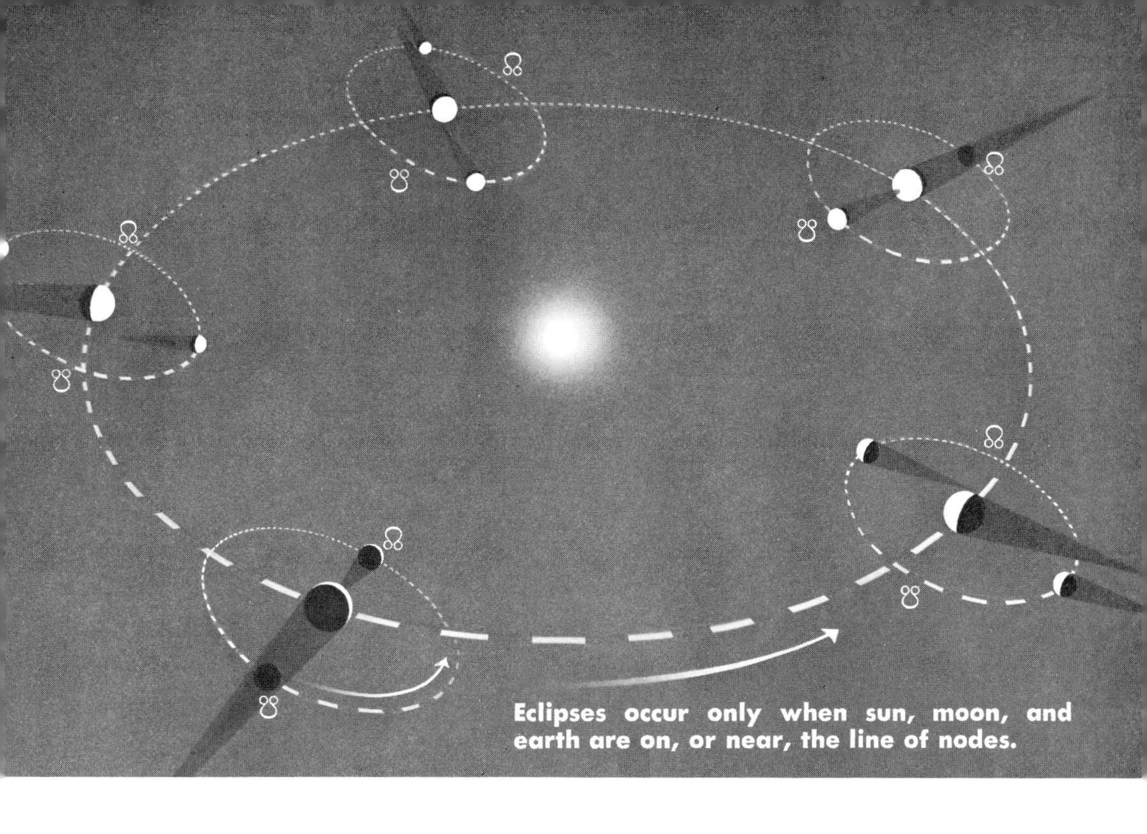
Eclipses occur only when sun, moon, and earth are on, or near, the line of nodes.

plane of earth's orbit during full moon, an eclipse occurs. Another way of saying the same thing is that a lunar eclipse occurs when the sun and moon appear near the nodes of moon's orbit. (The nodes are those places where the moon's orbit crosses the plane of earth's orbit. See the diagram.) When the moon moving northward crosses the plane of earth's orbit, it is at the ascending node. When the moon moving southward crosses the plane of earth's orbit, it is at the descending node. The ascending node is represented like this ☊; the descending node like this ☋.

If there were dust in space, so that space were not completely black, and if we could observe earth from a distant space platform, we would see earth's shadow. It extends some 860,000 miles into space. At the distance of the moon, about 238,000 miles, the shadow is about 5700 miles in diameter. Earth's shadow has a gray part and a dark part. The gray part is called the penumbra; the dark part is

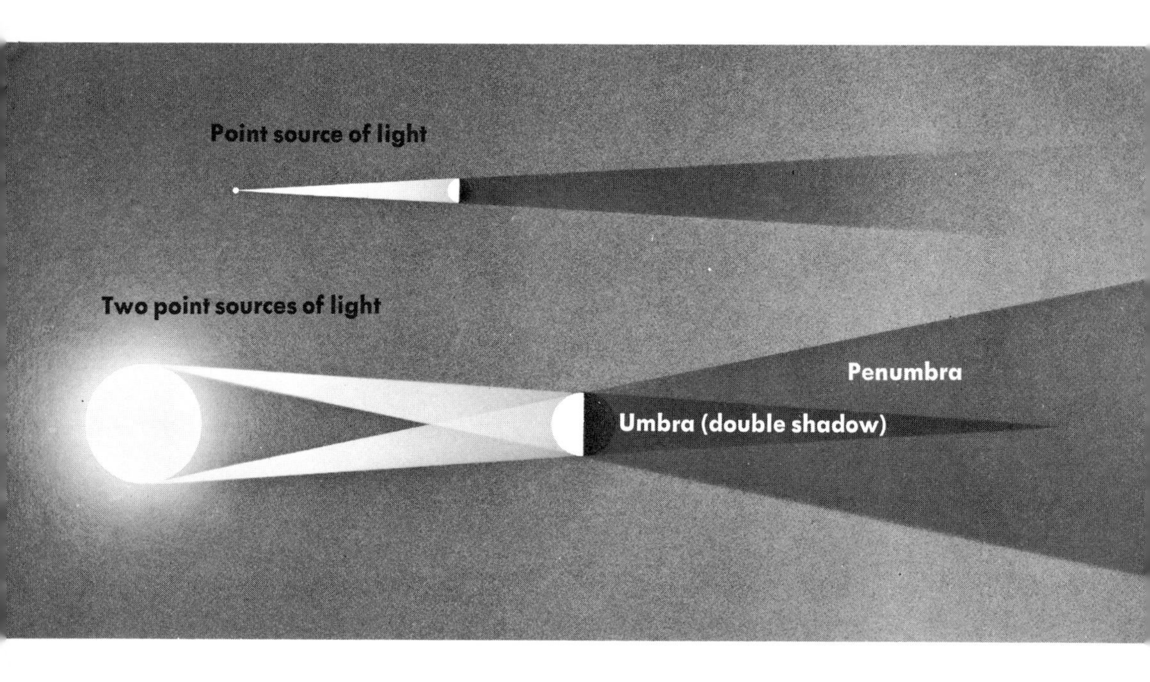


the umbra. The umbra is really a double shadow, as shown in the illustration.

It is impossible to notice when the moon enters the penumbra of earth's shadow, for the light is dimmed very little. When the moon enters the umbra, a dark region with a curved boundary is notched in the eastern edge of the moon. This happens because the moon moves into the shadow from the west. The moon moves deeper and deeper into the umbra; the notched portion increases in extent; and, after about one hour, the moon is completely in shadow. But it is not in darkness.

Even when the moon is in total eclipse, the moon is still visible from earth as a coppery-red disk. This is because earth's atmosphere bends, or refracts, sunlight and the red part mixes with earth's shadow.

Sunlight is made of light of many different colors—red, orange, yellow, green, blue, and violet. Each color corresponds to a particular wavelength. Red is a long-wave color, and as we move toward the violet, the wavelength decreases. When sunlight goes through our atmosphere, all the different waves are bent, or refracted. The blues are scattered by earth's atmosphere, and therefore they do not reach the moon. The reds are not scattered completely by the atmosphere, and some of them reach the moon.

Very rarely there is an eclipse when the moon is completely invisible, when there are so many clouds that the red light does not fall on the eclipsed moon. Such an eclipse occurred in 1884.

After totality, which occurs when the entire moon is in earth's shadow, the moon moves out of earth's shadow. Totality may last an hour or more. All the events described above happen once more. But this time they occur in reverse order.

Generally speaking, lunar eclipses are not nearly as valuable to astronomers as solar eclipses. However, ancient eclipses, once identified, can be used to determine the dates when events of long ago occurred. Astronomers can trace back eclipses, and fix quite precisely

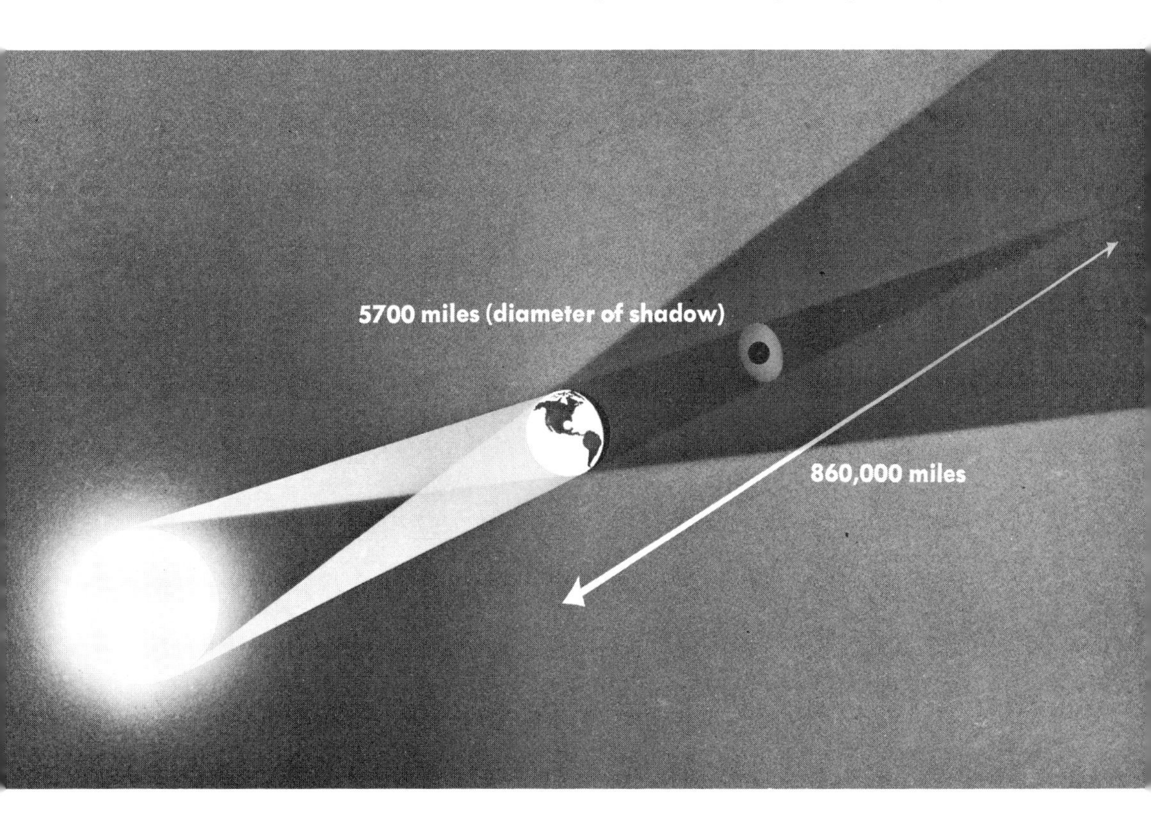

the dates when they took place. For example, we know from the writings of Josephus, an ancient Jewish historian, that there was a lunar eclipse the night before King Herod's death. This eclipse has helped historians fix the date of his death, and once this was known the date for the beginning of the Christian era could be determined without difficulty.

## Solar Eclipses

A solar eclipse occurs when the moon is between earth and sun, that is, when the moon is new. The length of the shadow cast by the moon varies between 228,000 and 236,000 miles. The average distance of the moon from the earth is 238,900 miles. Therefore, most of the time when the sun, earth, and moon are in the proper positions for an eclipse to occur, moon's shadow does not reach earth. However, the moon often is much closer to earth than the average, or mean, distance. Indeed, it may come within 217,750 miles of earth's surface. At such times it is possible for the shadow to extend more than 18,000 miles beyond earth, and for the shadow on earth to be some 167 miles in diameter. This would cause a total solar eclipse in that region of earth.

The moon may be 248,500 miles from earth's surface. The lunar shadow may be only 228,200 miles long. At such a time, the moon's shadow would fall some 20,000 miles short of reaching earth. An observer on earth and within the shadow cone looking toward the moon and sun would see a bright ring, or annulus, around the moon. The bright ring is a ring of sunlight. This is an annular eclipse, an event that is more common than a total solar eclipse.

An observer anywhere within the gray part of the moon's shadow, the penumbra, will see a partial eclipse of the sun. That is, if he is near the umbra, he will see only a sliver of the sun. If he is far from the umbra, up to about 2000 miles, he will see the moon cover only a small notch of the solar disk. Because the area of partial eclipse is so much broader than the narrow belt of totality, partial eclipses are seen much more frequently at any given location than are total

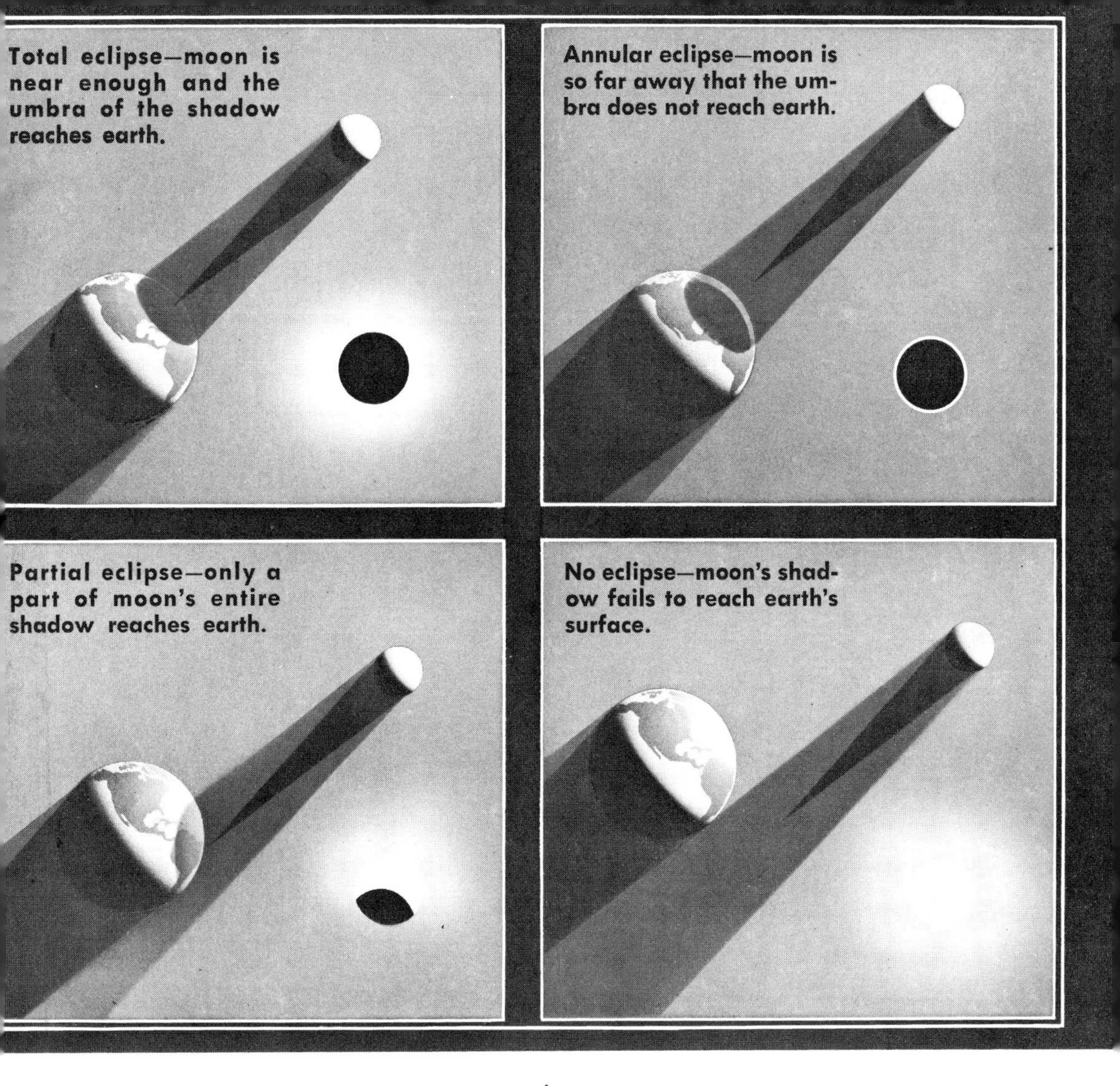


eclipses. For example, the last total eclipse visible from New York City was in 1925, the next total eclipse visible from the same location will occur in the year 2024. During this interval of 99 years, several partial eclipses have been and will be seen from New York City.

Astronomers who study the sun during a total eclipse must plan carefully ahead of time, for the event lasts only a few minutes. They must test and retest their equipment, be sure all the cameras are working, and know precisely what they are going to do when the eclipse begins. The longest that an eclipse can last is 7 minutes and 40 seconds. It can last this long only when viewed from the equator. Farther north, the maximum time of totality becomes less. Around the vicinity of Boston an eclipse can last no longer than about 6½ minutes.

Astronomers time eclipses by four separate contacts. The first contact occurs when the edge of the moon first touches the edge of the sun. Second contact occurs when the eclipse becomes total. Third contact occurs when the moon begins to move beyond the sun, and fourth contact is the moment that the eclipse is over. This series of contacts may take 4 hours from start to finish.

A total eclipse of the sun is a remarkable sight indeed. No wonder the ancient peoples who were ignorant of the causes of an eclipse were

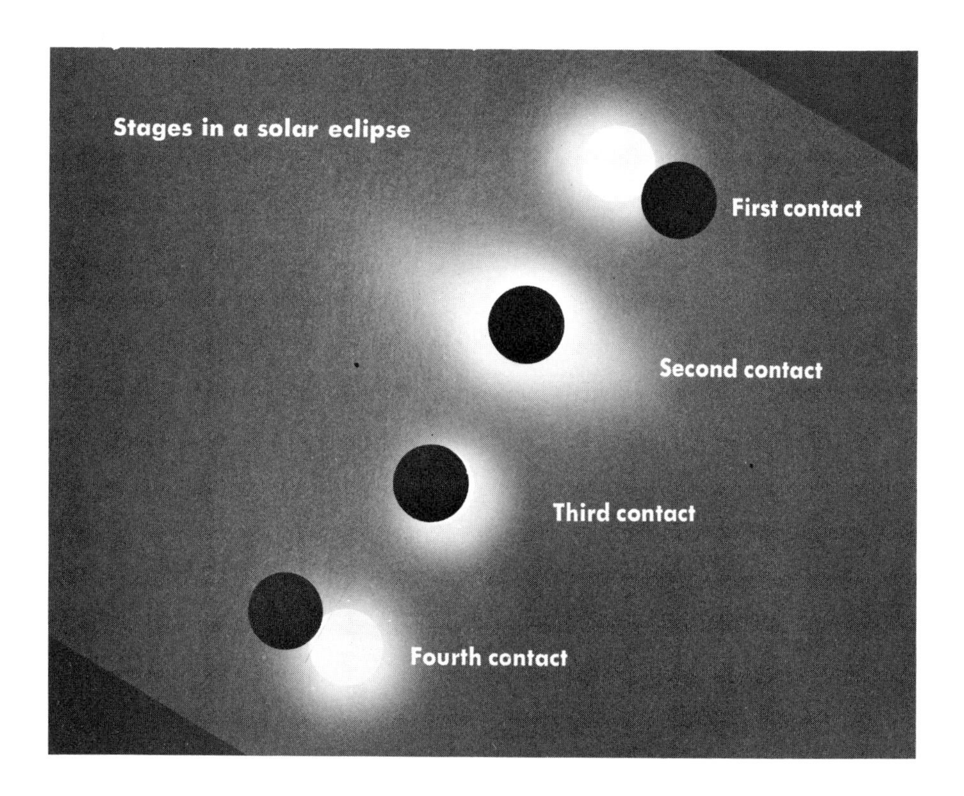

Baily's beads and the diamond-ring effect

struck with great fear as the skies darkened. Even today, there are many tribes that are awestruck by a solar eclipse.

Not surprising! A few minutes before totality, darkness begins to spread across the countryside. Whatever light is present comes from only the edge of the sun, so that the sunlight loses much of its intensity. Chickens go to roost, apparently believing night has come. Just as totality nears, people who are viewing the eclipse from a high location can see the moon's shadow speeding across the countryside.

An instant before totality, Baily's beads become visible. This is a string of bright lights along the edge of the moon's disk that were first explained by Baily, an English astronomer. The bright lights are produced by sunlight shining through the deep valleys between high mountains on the surface of the moon. The lights appear as a flash, and then the sun is in total eclipse; the moon blots out the disk completely.

The moon is able to obscure the sun because of a strange set of circumstances. The diameter of the moon is 2160 miles, and the diameter of the sun is 864,000 miles—some 400 times greater. But the *distance* from earth to sun is about 400 times greater than the distance from earth to moon, 93,000,000 miles to the sun against 240,000 miles to the moon.

Because the moon is 400 times closer to earth than is the sun, and because the sun's diameter is 400 times greater than the moon's diameter, the much smaller moon can obscure the sun.

You can get the effect by holding a marble with your thumb and first finger at arm's length. Look at the marble, and also at a lamp across the room. Keep looking at the marble as you move it closer to your eye. Soon the lamp will be obscured; only the marble will be visible. The glow of the lamplight can be seen all around the marble.

Astronomers understand the motions of the sun, earth, and moon so completely that they can predict precisely the relative locations of these three objects as they were centuries ago, and as they will be in the future. Much of the work of an astronomer is concerned with meticulous mathematical computations. A German astronomer-mathematician named Oppolzer computed the occurrence and the coverage of eclipses for a period of over twenty centuries. It is likely that Oppolzer saw only half a dozen eclipses during his lifetime, yet his ingenuity and knowledge gave him considerable information about events that had taken place long before he was born and that would occur long after his death.

# The Moon's Surface

Take a careful look at the moon when it is full. You will see bright and dark zones, the arrangement of which reminds you of a man's eyes, nose, and mouth. In 1609 Galileo, the famous Italian astronomer, studied the moon with his homemade telescope and saw that the "man in the moon" was produced by dark regions separated by lighter areas.

Since the time of Galileo, extensive study of the moon has been made, and very fine photographs of the moon have been produced. The picture on page 44 is a composite of two photographs. It shows a full moon. In maps used in astronomy, unlike those used for the earth, the directions are reversed: south is at the top and north is at the bottom. But in such a reversed map, east is at the right and west at the left. There are good reasons why this is so.

A telescope produces an upside-down, or inverted, image. To right the image, as in an ordinary terrestrial telescope, another glass called a correcting lens would be necessary. Every time that light goes through a lens, some of the light is lost. Therefore, the image that results is less sharp. Rather than make such a sacrifice, and in order to collect as much light as possible, astronomers do not use a correcting lens. East and west appear at the right and left because, when you view the moon with the naked eye, the western edge is at your right.

In reality, the full moon never looks like the photograph shown here. The various features do not stand out sharply during full moon. As a matter of fact, the lower features cannot be observed during full moon because they are lighted from above and there are, therefore, no shadows to define boundaries. Indeed, shadows are very important, for they enable us to define the craters, mountains, and peaks. The craters are seen best when they are at or near the terminator, the line that separates the lighted half of the moon from the unlighted half, for then the shadows are the longest.

The terminator, the line that separates daylight from darkness, sweeps across the entire lunar surface twice a month. First it reveals a thin sliver of the moon. At full moon there is no terminator. Then the terminator sweeps across the surface, reducing the lighted area.

When a crater is near the terminator, the crater is lighted from the side. Therefore, the crater walls cast deep shadows. The shadows contrast with the bright lunar surface, and the craters are clearly visible.

When one first observes the moon, the surface appears to be a mixture of various formations. On closer investigation, the observer finds that there are distinct bright areas, called craters, and dark areas called mares, or seas.

Many of the more prominent craters can be identified in the drawings which appear on pages 54 to 59.

Galileo was the first to call the great lunar plains *maria*, or seas. He believed that these great flat areas that were so clearly seen through his small telescope might be filled with water.

Riccioli, an Italian astronomer, made the first moon map in 1650. He had to have some way of identifying the various formations, and he therefore devised the following procedure.

There are ten mountain ranges on the moon. These are named after mountain ranges on earth, such as Alps, Apennines, Caucasian. However, a few carry the names of astronomers, such as Doerfel and Leibnitz.

The large, clearly visible craters are named after ancient astronomers and philosophers. They are named Archimedes, Copernicus, Kepler, Plato, Tycho. There are some 32,000 craters on the visible side of the moon, and a great many of them are named after more modern astronomers and lunar observers.

The word *crater* implies some sort of hole or depression formed by a volcano. The term is unfortunate, for many of the craters of the moon might more properly be called walled plains because they are so extensive. For example, some craters are 50 to 60 miles across, and the largest one visible at all librations is Clavius—160 miles.

The inner walls of most of the moon's craters are very high; often they reach 5000 feet; some of them attain 17,000 feet; and the inside walls of Newton, the deepest crater, rise to some 29,000 feet. However, the outer walls rise only a thousand feet or so above the surrounding region. Probably all the craters are circular. Look closely at the picture on page 44; the craters near the edge, or limb, of the moon appear to be elliptical. However, this is due only to the angle at which we view them. If we could see these craters from overhead, they would be circular.

The largest craters are probably the oldest. Plato, which is on the northern edge of Mare Imbrium, appears to have a smooth floor. We believe it is an old crater because it is rather shallow; it may have been deeper at some time but has become filled with dust and other materials. At the arrow on the picture on page 44 are craters of three different ages. The craters are Ptolemaeus, Alphonsus, and Arzachel. If you have a pair of 7 × 50 binoculars you might be able to see these craters when the moon is just beyond first quarter. Mount the binoculars on a tripod so they do not jiggle. Look carefully at the picture on page 77. Notice that Ptolemaeus is not very deep, Alphonsus is a bit deeper, and Arzachel is deepest of all. Some astronomers believe that craters gradually become filled with dust that falls upon the moon from outer space. Others believe that the dust is formed when the rocks of the lunar surface break up. The intense heating and cooling of the surface causes the rocks to crumble to dust. The dust sifts down to the lowest level, and the craters become filled with hundreds, even thousands, of feet of lunar debris.

Notice that the floor of Arzachel is rough. Small craterlets still protrude from it. Over the centuries these mounds may become covered with dust, and finally they may disappear from view completely.

The Mare Imbrium area of the moon is especially interesting. Early observers of the moon noticed that some regions were smooth and flat. They called these regions seas. The term is not a good one, for there are no seas on the moon. Also, the regions are probably not nearly as smooth as they appear to be when viewed through a low-power telescope. They may be filled with crevices and rough areas, as modern instruments indicate.

The "level" regions are given poetic names. Mare Imbrium means the Sea of Showers; Mare Frigoris, the Sea of Cold; Mare Serenitatis, the Sea of Serenity; Mare Crisium, the Sea of Crises; Mare Tranquillitatis, the Sea of Tranquillity.

Mare Imbrium is an extensive plain some 750 miles across. It is

bounded by high mountain ranges: the Carpathians to the southeast, the Apennines to the southwest, the Caucasian and the Alps to the northwest, and the Jura Mountains to the northeast. The Jura Mountains wall in the Sinus Iridum, the Bay of Rainbows. This is a large, flat region which long ago may have been a mare itself.

Toward the northwest of Mare Imbrium is Plato, a shallow, old crater. Notice how smooth the floor of the crater is. Almost directly south of Plato is Archimedes, another crater having a smooth floor.

Toward the east, Mare Imbrium blends into Oceanus Procellarum, the Ocean of Storms. The crater Copernicus dominates this region. This is especially true during full moon, when white rays can be seen radiating from the crater. We do not know the nature of these rays. They may be made of ashes that were thrown out by volcanoes which exploded as they came to the end of their activity.

Ever since Galileo first observed the many different formations on the moon, men have wondered how they were made. Photographs of the surface, such as those made by the Ranger and Surveyor probes and by astronauts, provide clues which indicate that at least some of the craters were formed by volcanic activity. Many of the craters on the moon are scores of times larger than we believe a volcanic crater can be. Some astronomers maintain that a volcano on the moon would produce a very large crater because of moon's low gravitational attraction. But other astronomers maintain that it is unlikely that craters 50 to 60 miles across could have been produced by volcanic action of the kind we have on earth.

Another theory proposes that the craters of the moon were produced by collisions with meteorites. Meteorites may have collided with the moon long ago, when this neighbor world of ours was in a plastic condition. The meteorites would have been very large and very numerous. They could have been remnants of materials that did not join together to make the moon. Very likely the meteorites struck the lunar surface at high speeds, so high that a tremendous amount of heat was generated on impact. The heat vaporized the meteorite, so that tremendous pressure built up rapidly. The pressure may have been sufficient to gouge out the craters.

Left: Surface features as small as 75 feet across can be seen in this photograph made by Lunar Orbiter V.

Below: Oblique view looking northwest into the Sea of Tranquillity. Photograph was taken from the Apollo 8 spacecraft.

The crater Hyginus is at the center, and two rilles (valleys) extend from it in this photograph made by Lunar Orbiter III.

The crater Goclenius photographed by Apollo 8. The crater is about 40 miles in diameter and has a rille cutting through the rim.

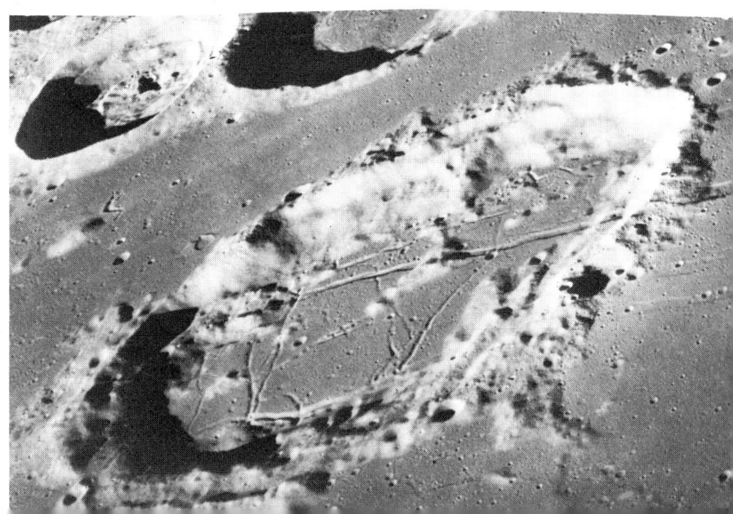

**This is the earliest photograph of the near side of the moon. It was made in 1852 by John Adams Whipple at the Harvard College Observatory.**

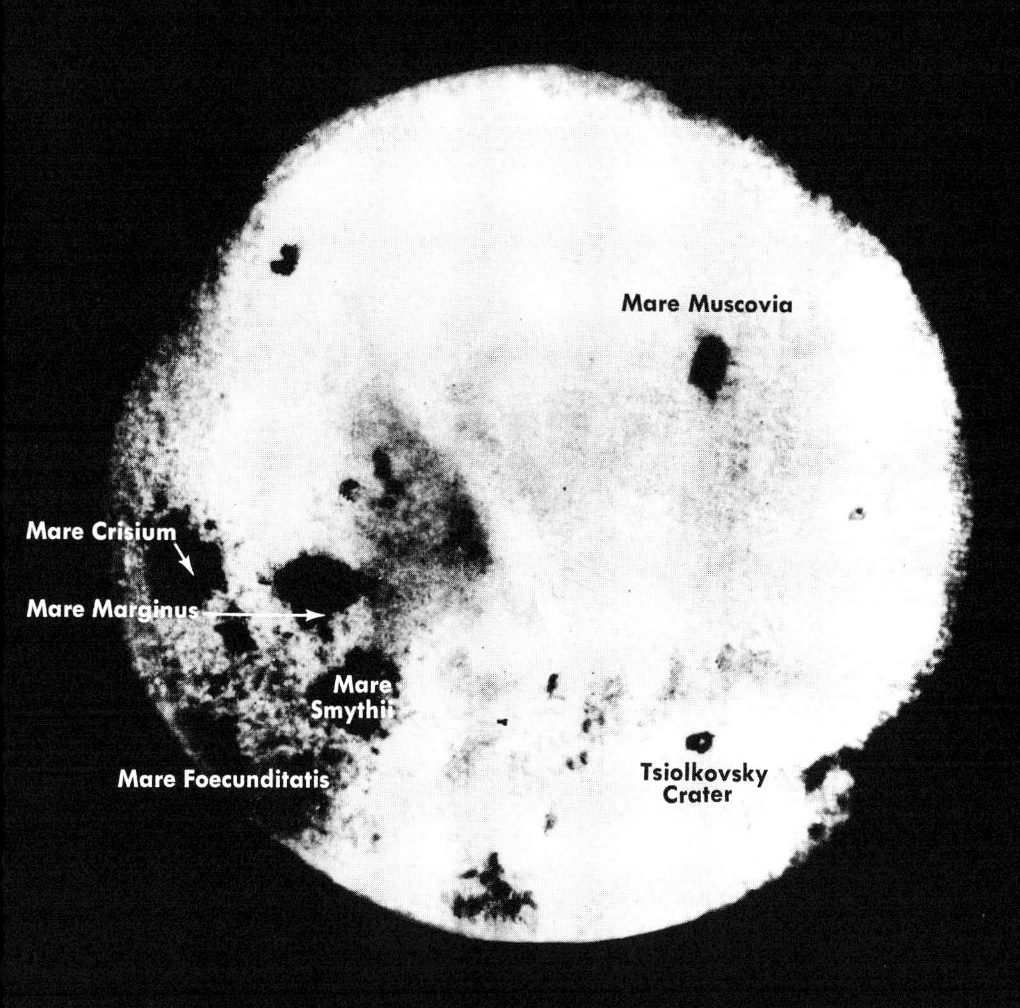


**This is the first photograph of the half of the moon that is always away from the earth. It was made in 1959 by Lunik III, a Russian satellite. *(Sovfoto)***

There is a strong possibility that earth's surface was scarred at one time with similar craters. However, unlike the moon, earth has atmosphere and water, two substances that wear down the surface. Through hundreds of thousands of years, the action of wind, rain, snow, and rivers would have worn down high crater walls and would have filled in the low crater floors.

In some respects we know more about the surface of the moon than we do about the surface of earth, for we can see so much of the lunar surface at one time. Painstaking hours of observing by multitudes of astronomers in all parts of the world, and down through the centuries since the invention of the telescope, have contributed to the vast store of knowledge now assembled. Our knowledge of the moon has been increased tremendously by the study of the moon rocks and soil brought back by astronauts of Apollo missions.

Careful study of the rocks by scores of scientists around the world revealed no signs of life on the moon presently, or at any time in the past. The rocks were collected from only three limited locations: a small part of the Sea of Tranquillity, the Ocean of Storms, and Fra Mauro. The rock was mainly basalt, a volcanic rock, and breccia, a formation made of small pieces of rock fused together. Many scientists believe the breccias were formed by intensive heating produced when meteorites struck the moon.

The minerals in the lunar rocks were mostly feldspar, pyroxene, and ilmenite. All these materials are found in earth rocks. However, the amount of ilmenite in the moon rocks was much greater than that found in earth rocks. Ilmenite is a dark mineral that is an ore of titanium. Perhaps it is this mineral that gives the dark coloring to the mares of the moon.

Most of the lunar rocks turned out to be 3.6 to 3.9 billion years old. However, one rock from the Ocean of Storms and the lunar soil was found to be even older—some 4.4 billion years old. This is believed to be the approximate age of the earth.

We are learning more and more about the moon, but we are still ignorant about many details. Even though men have walked on the moon, and even though we have detailed photographs of many areas of the moon, we still do not know for sure how craters were formed. However, it seems that they were produced by volcanic activity and by the impact of collisions with meteorites of great size. The presence of large amounts of titanium in lunar rocks is another puzzle. The moon still harbors many mysteries—where did it come from, what has been its history?

The drawings on the next few pages show the features on the near and far sides of the moon, as well as those in the north and south polar regions. Some of the lunar features are named in the accompanying table and may be located by finding the intersections of the lines indicated.

| *Lunar Feature* | *From Bottom to Top* | *From Left to Right* |
|---|---|---|
| Abulfeda | 13S | 13E |
| Albategnius | 11S | 4E |
| Alphonsus | 13S | 3W |
| Alter | 19N | 108W |
| Apollo | 37S | 153W |
| Archimedes | 30N | 4W |
| Aristarchus | 23N | 48W |
| Aristillus | 34N | 1E |
| Aristoteles | 50N | 17E |
| Arrhenius | 55S | 91W |
| Arzachel | 18S | 2W |
| Atlas | 47N | 44E |
| Autolycus | 31N | 1E |
| Avogadro | 64N | 165E |
| Becquerel | 41N | 129E |
| Belkovich | 62N | 90E |
| Bell | 22N | 96W |
| Berkner | 25N | 105W |
| Birkhoff | 59N | 148W |
| Borman | 39S | 149W |
| Brunner | 10S | 91E |
| Byrd | 85N | 2E |

# LUNAR EARTHSIDE CHART

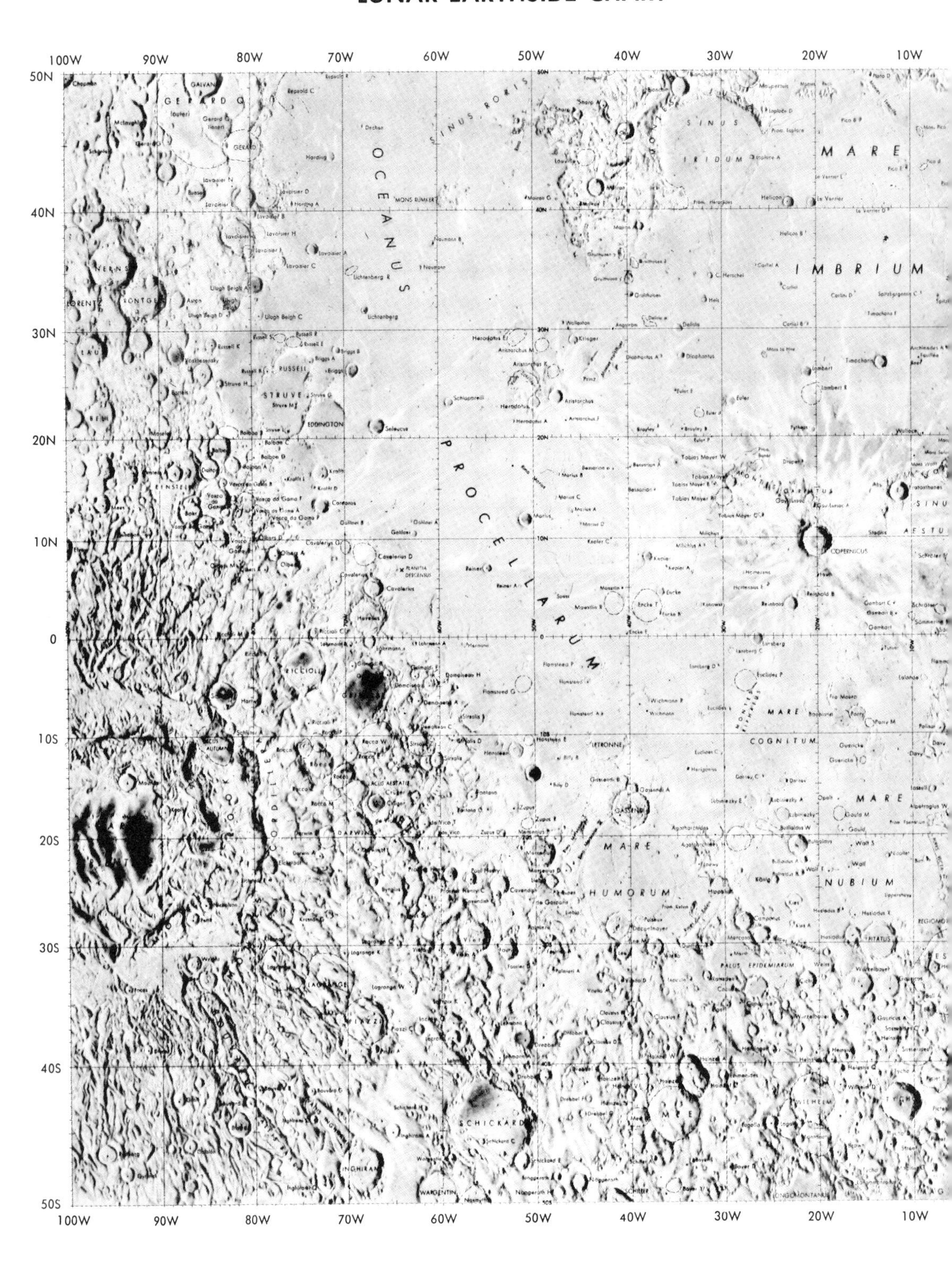

# LUNAR EARTHSIDE CHART

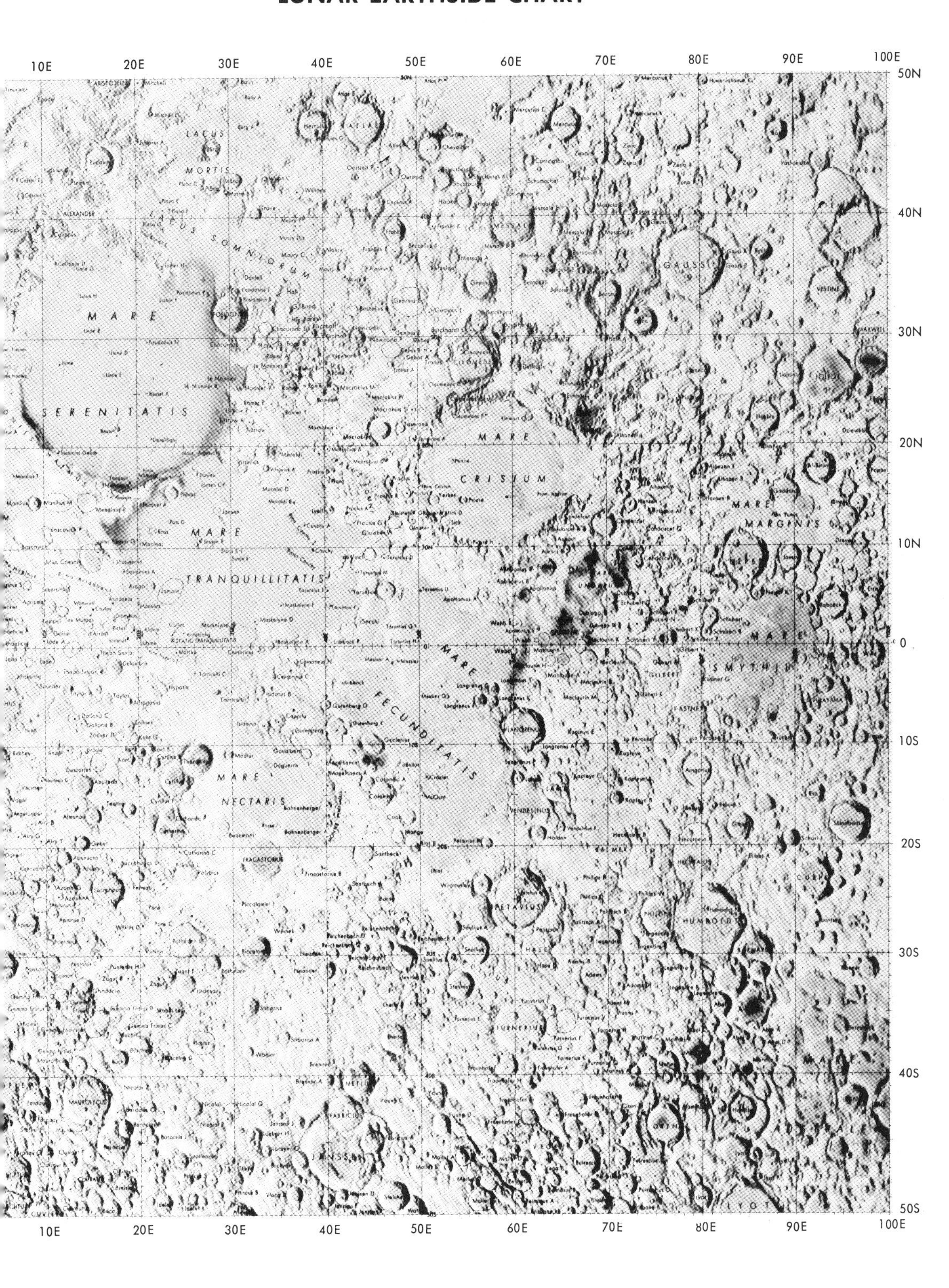

# LUNAR FARSIDE CHART

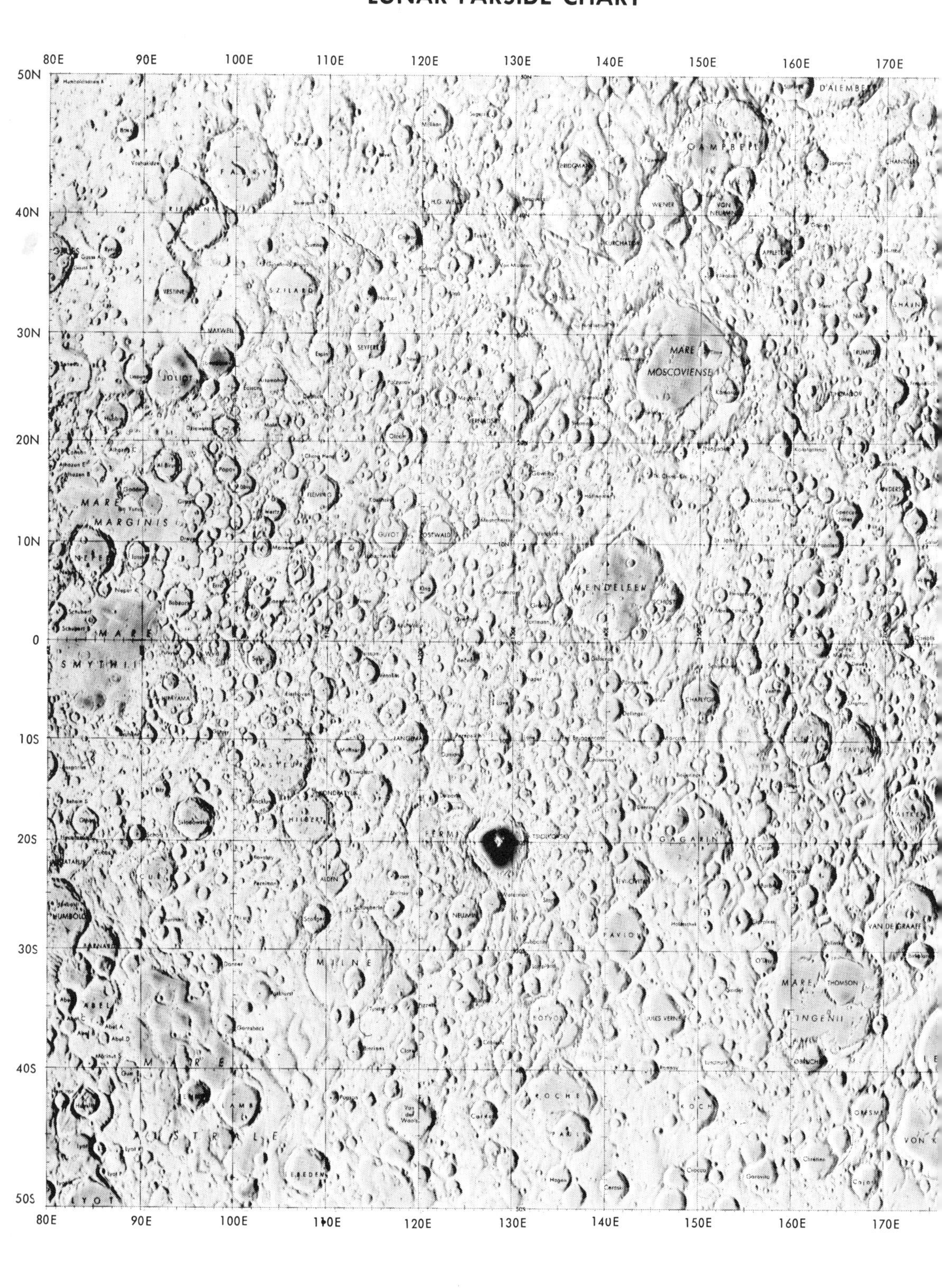

# LUNAR FARSIDE CHART

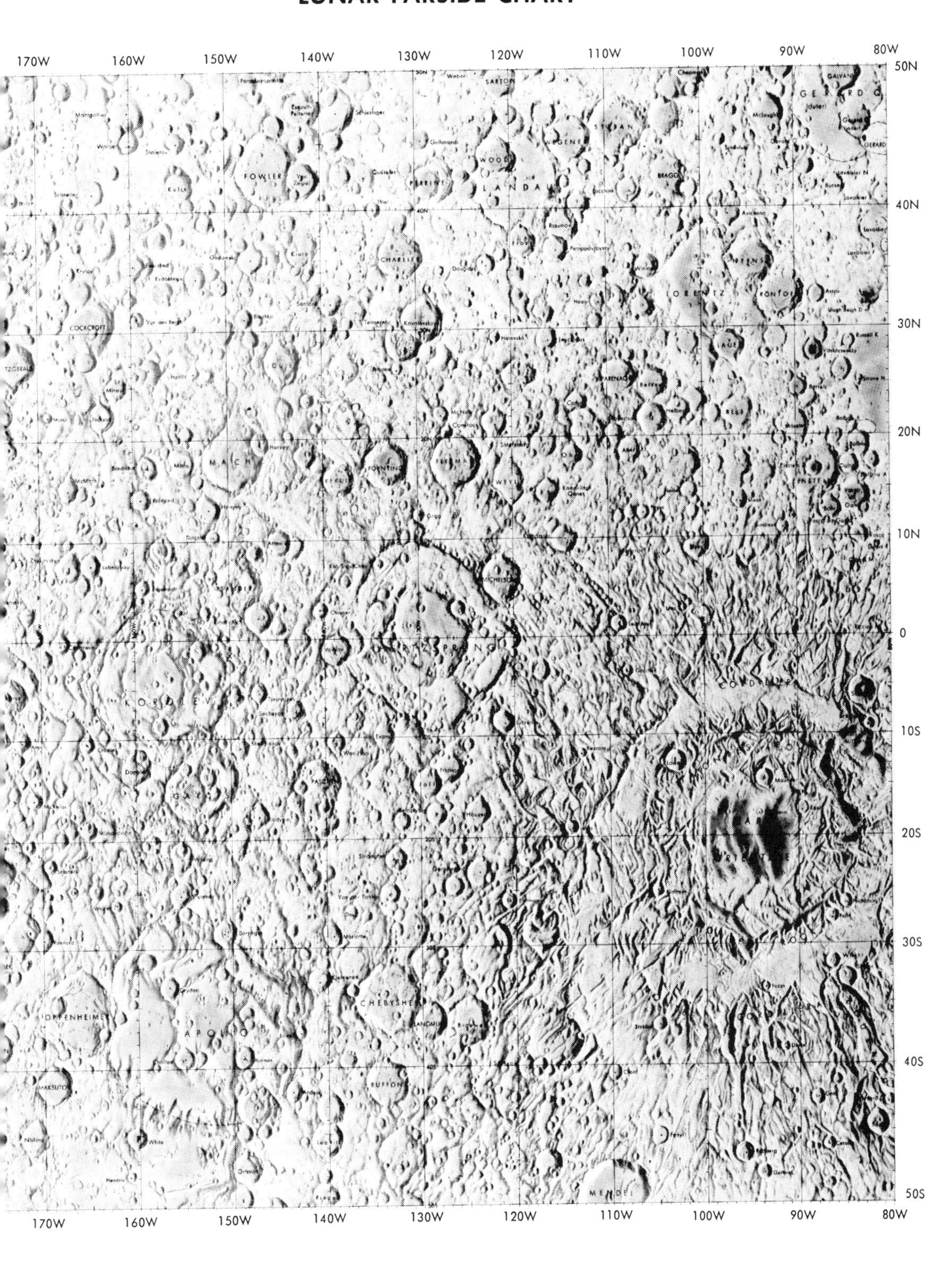

# LUNAR POLAR CHART

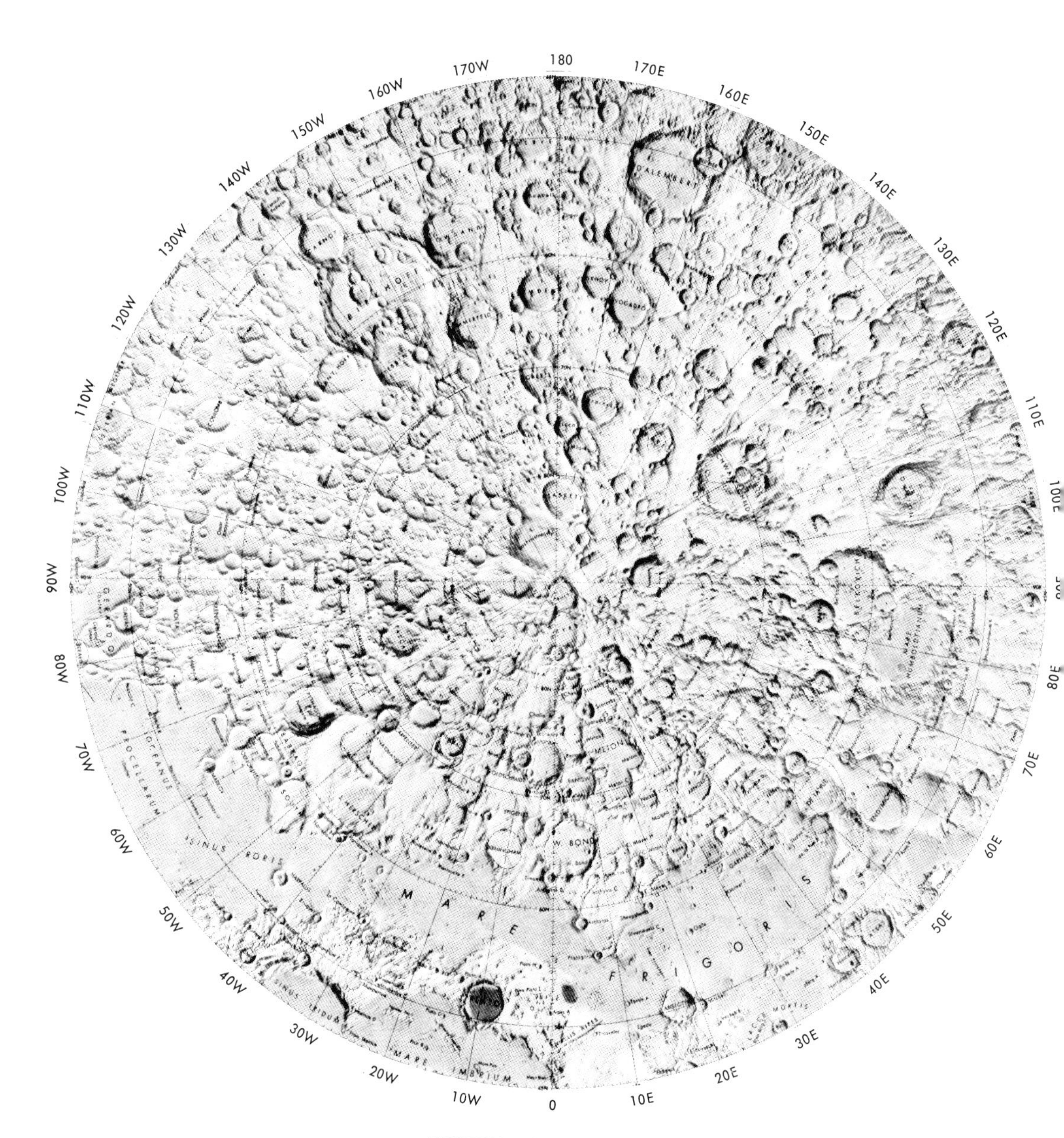


## NORTH POLAR REGION

# LUNAR POLAR CHART

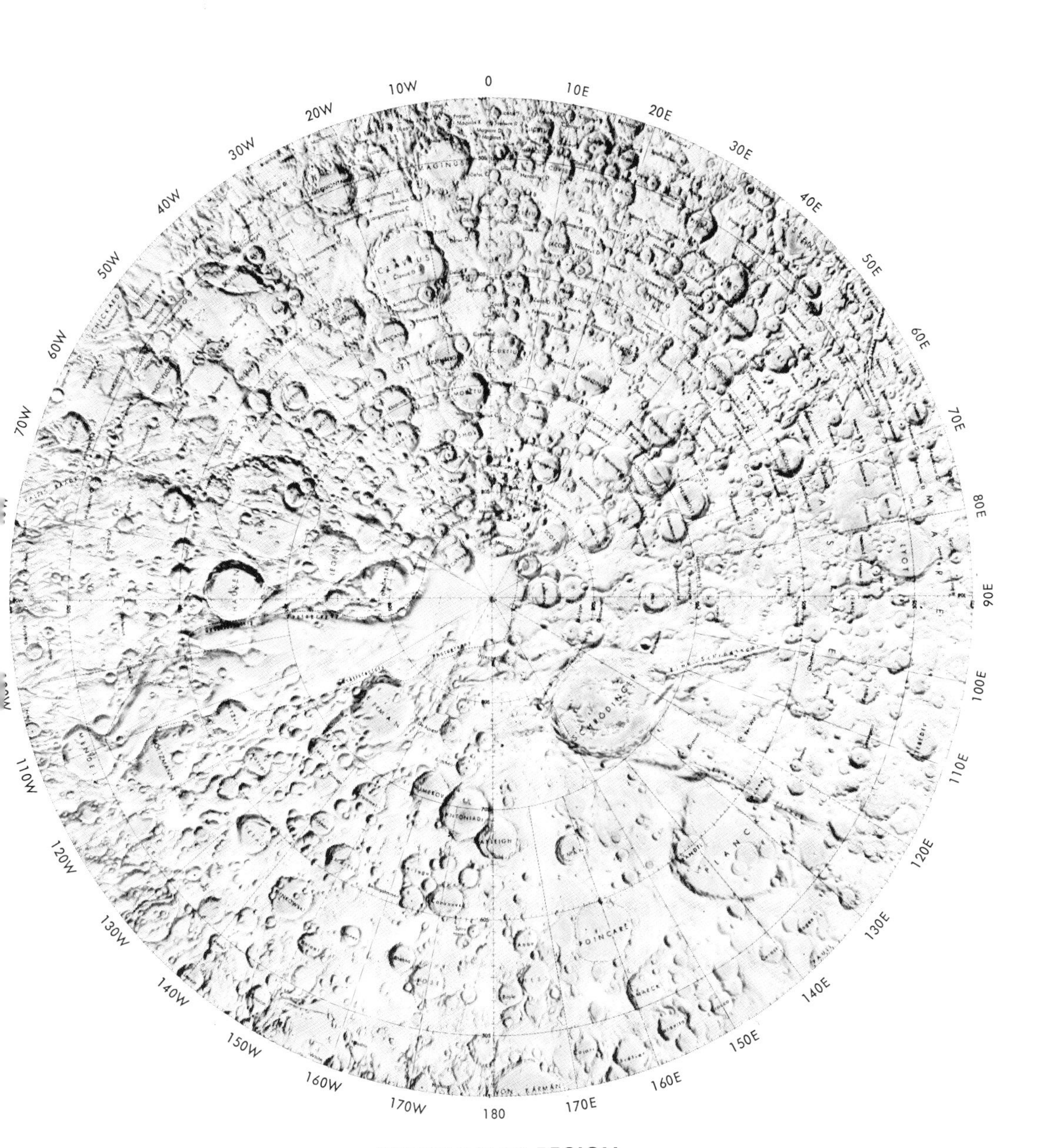


SOUTH POLAR REGION

| *Lunar Feature* | *From Bottom to Top* | *From Left to Right* |
|---|---|---|
| Campbell | 45N | 152E |
| Carnot | 52N | 144W |
| Cassegrain | 52S | 113E |
| Clavius | 58S | 15W |
| Cleomedes | 28N | 56E |
| Coblentz | 38S | 126E |
| Compton | 57N | 105E |
| Comstock | 21N | 122W |
| Copernicus | 10N | 20W |
| Curie | 23S | 92E |
| Daedalus | 6S | 180 |
| D'Alembert | 52N | 164E |
| Darwin | 20S | 70W |
| De La Rue | 59N | 52E |
| Deslandres | 31S | 5W |
| Dewar | 3S | 166E |
| Doppelmayer | 28S | 41W |
| Doppler | 13S | 160W |
| Dreyer | 10N | 97E |
| Ehrlich | 41N | 172W |
| Einstein | 15N | 89W |
| Endymion | 53N | 55E |
| Engelhardt | 5N | 159W |
| Eratosthenes | 15N | 12W |
| Eudoxus | 44N | 16E |
| Evans | 10S | 134W |
| Fabry | 43N | 100E |
| Fermi | 20S | 112E |
| Fizeau | 58S | 133W |
| Fleming | 15N | 109E |
| Fowler | 43N | 145W |
| Fracastorius | 21S | 33E |
| Furnerius | 36S | 60E |
| Gagarin | 20S | 150E |
| Gassendi | 17S | 40W |

| *Lunar Feature* | *From Bottom to Top* | *From Left to Right* |
|---|---|---|
| Gauss | 35N | 78E |
| Geiger | 14S | 158E |
| Gerard Q | 48N | 85W |
| Goclenius | 10S | 45E |
| Green | 4N | 133E |
| Grimaldi | 6S | 68W |
| Grissom | 48S | 149W |
| Harriot | 33N | 114E |
| Hausen | 65S | 90W |
| Heaviside | 10S | 167E |
| Hell | 32S | 8W |
| Hercules | 47N | 39E |
| Herodotus | 23N | 50W |
| Hertzsprung | 0 | 130W |
| Hipparchus | 5S | 5E |
| Hippocrates | 71N | 146W |
| Hirayama | 6S | 93E |
| Humboldt | 27S | 81E |
| Hyginus | 8N | 6E |
| Icarus | 6S | 173W |
| Jackson | 22N | 163W |
| Janssen | 46S | 42E |
| Jenner | 42S | 96E |
| Joliot | 26N | 93E |
| Karpinsky | 73N | 166E |
| King | 5N | 120E |
| Komarov | 25N | 153E |
| Korolev | 5S | 157W |
| Lampland | 31S | 131E |
| Landau | 42N | 119W |
| Lacus Mortis | 45N | 27E |
| Lacus Somniorum | 38N | 30E |
| Lambert | 26N | 21W |
| Langmuir | 36S | 129W |
| Langrenus | 8S | 61E |

| *Lunar Feature* | *From Bottom to Top* | *From Left to Right* |
|---|---|---|
| Laue | 28N | 97W |
| Leavitt | 46S | 140W |
| Letronne | 11S | 42W |
| Lindblad | 70N | 99W |
| Longomontanus | 50S | 22W |
| Lovell | 37S | 142W |
| Lowell | 13S | 103W |
| Lyot | 50S | 85E |
| Mach | 18N | 149W |
| Maginus | 50S | 6W |
| Maksutov | 41S | 169W |
| Mare Australe | 40S | 95E |
| Mare Cognitum | 9S | 24W |
| Mare Crisium | 17N | 60E |
| Mare Fecunditatis | 8S | 52E |
| Mare Frigoris | 55N | 10E |
| Mare Humboldtianum | 56N | 80E |
| Mare Humorum | 23S | 39W |
| Mare Imbrium | 35N | 25W |
| Mare Ingenii | 34S | 163E |
| Mare Marginis | 13N | 88E |
| Mare Nectaris | 15S | 32E |
| Mare Nubium | 20S | 15W |
| Mare Orientale | 20S | 95W |
| Mare Serenitatis | 30N | 20E |
| Mare Smythii | 1N | 85E |
| Mare Spumans | 2N | 65E |
| Mare Tranquillitatis | 10N | 30E |
| Mare Undarum | 8N | 68E |
| Mare Vaporum | 15N | 2E |
| Marius | 12N | 51W |
| McLaughlin | 47N | 93W |
| McMath | 15N | 167W |
| Mee | 44S | 36W |
| Meitner | 11S | 113E |

| *Lunar Feature* | *From Bottom to Top* | *From Left to Right* |
|---|---|---|
| Mendeleev | 6N | 141E |
| Mercurius | 47N | 66E |
| Merrill | 75N | 116W |
| Messala | 39N | 60E |
| Meton | 74N | 20E |
| Michelson | 6N | 121W |
| Millikan | 47N | 121E |
| Minkowski | 56S | 145W |
| Montes Alpes | 47N | 1W |
| Montes Apenninus | 20N | 2W |
| Montes Carpatus | 15N | 23W |
| Montes Caucasus | 36N | 9E |
| Montes Cordillera | 20S | 80W |
| | 35S | 100W |
| Montes Haemus | 20N | 10E |
| Montes Jura | 48N | 33W |
| Montes Pyrenaeus | 16S | 41E |
| Montes Rook | 12S | 90W |
| | 30S | 90W |
| Neper | 8N | 85E |
| Nernst | 36N | 95W |
| Nobel | 15N | 101W |
| Oceanus Procellarum | 15N | 60W |
| O'Day | 31S | 157E |
| Oken | 44S | 76E |
| Omar Khayyam | 58N | 102W |
| Oppenheimer | 35S | 166W |
| Palus Putredinis | 28N | 1W |
| Palus Somni | 15N | 42E |
| Pannekoek | 4S | 141E |
| Paracelsus | 23S | 163E |
| Pasteur | 12S | 105E |
| Peary | 89N | 30E |
| Perkin | 47N | 176W |
| Petavius | 25S | 61E |

| *Lunar Feature* | *From Bottom to Top* | *From Left to Right* |
|---|---|---|
| Pitatus | 30S | 14W |
| Planck | 57S | 135E |
| Plaskett | 82N | 175E |
| Plato | 51N | 10W |
| Posidonius | 32N | 30E |
| Ptolemaeus | 9S | 2W |
| Purbach | 26S | 2W |
| Pythagoras | 64N | 62W |
| Riccioli | 3S | 74W |
| Riemann | 40N | 96E |
| Rittenhouse | 74S | 107E |
| Roche | 42S | 135E |
| Röntgen | 32N | 92W |
| Rumford | 29S | 170W |
| Schickard | 45S | 55W |
| Schiller | 52S | 40W |
| Schrödinger | 75S | 133E |
| Schwarzschild | 71N | 120E |
| Seleucus | 21N | 66W |
| Seyfert | 29N | 114E |
| Sinus Aestuum | 12N | 7W |
| Sinus Iridum | 45N | 31W |
| Sinus Medii | 1N | 1W |
| Sinus Roris | 50N | 55W |
| Sklodowska | 18S | 96E |
| Spencer Jones | 13N | 166E |
| Stebbins | 65N | 143W |
| Stofler | 41S | 5E |
| Strömgren | 22S | 133W |
| Struve | 24N | 78W |
| Szilard | 34N | 106E |
| Taruntius | 5N | 46E |
| Tereshkova | 28N | 145E |
| Tesla | 38N | 125E |
| Theophilus | 12S | 26E |

| *Lunar Feature* | *From Bottom to Top* | *From Left to Right* |
|---|---|---|
| Timocharis | 27N | 13W |
| Titov | 28N | 150E |
| Tycho | 43S | 11W |
| Tyndall | 35S | 117E |
| Vallis Alpes | 49N | 2E |
| Vallis Rheita | 44S | 52E |
| Vallis Snellius | 30S | 51E |
| Vasco da Gama | 14N | 84W |
| Vendelinus | 17S | 61E |
| Vestine | 34N | 93E |
| Vitello | 30S | 38W |
| Von Neumann | 40N | 153E |
| Walter | 33S | 1W |
| Wan-Hoo | 11S | 139W |
| Wells, H. G. | 41N | 122E |
| Wilhelm | 43S | 21W |
| Xenophanes | 58N | 83W |
| Zeeman | 75S | 135W |
| Zhukovsky | 7N | 167W |

# Temperatures on the Moon

Sometimes it is very hot on the surface of the moon; the temperature exceeds the boiling point of water here on earth. At other times the temperature is very low; it is much colder than any place on the surface of earth.

At the subsolar point on the moon, the location where the sun would appear directly overhead to an observer located on the surface, the temperature soars to 240° F. above zero. As one moves away from the subsolar point, the temperature drops off rapidly, reaching 58° F. below zero at those places where the sun would appear to our imaginary observer to be on the lunar horizon.

The temperature readings at the subsolar point are not always the same. For example, at the subsolar point at full moon the temperature may reach 240° F., but only 180° F. during first quarter. Apparently the difference is caused by variations in the lunar surface; a rough surface increases the reflection of heat to our measuring instruments, so that the temperature readings are higher.

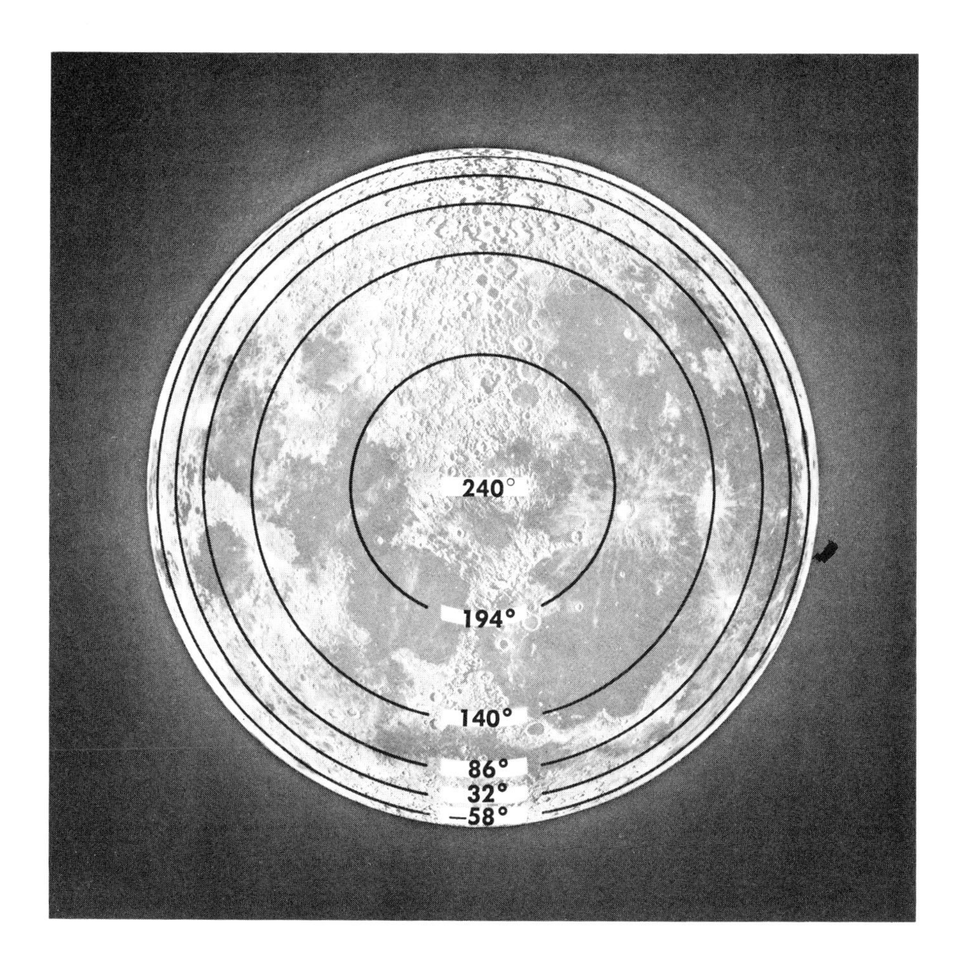


Temperature measurements have been made on the dark side of the moon. Astronomers throughout the world would like to know the correct temperatures, and therefore after they make observations they publish their findings. Other astronomers make observations to see if their results agree with the earlier ones. Usually the results do not agree precisely, but the astronomers can get a general idea of temperatures by averaging the many different readings. The average temperature of the dark side of the moon is 270° F. below zero. But such a low reading cannot be relied upon, because the measuring

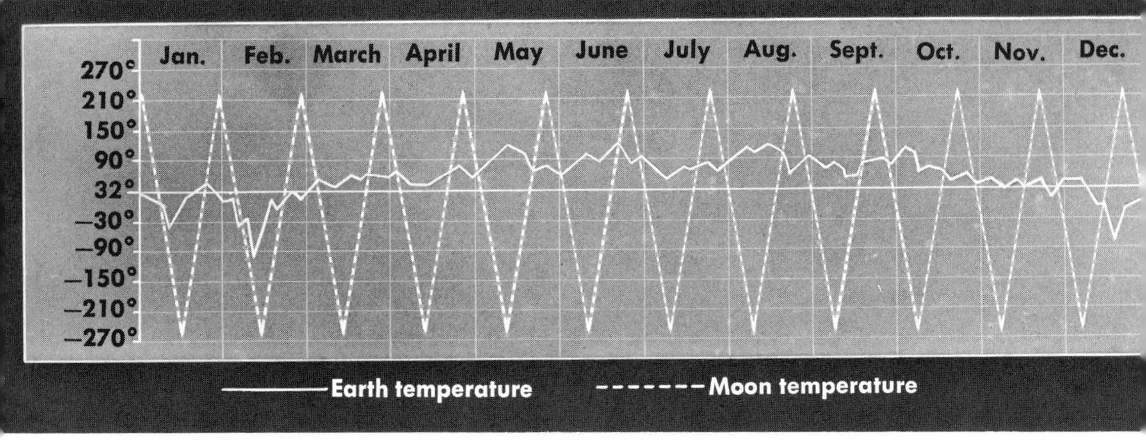


devices that are now available are not especially sensitive in this low range.

Measuring higher ranges of lunar temperatures has become quite accurate since men have landed there and so have been able to place thermometers on the surface.

The range of temperatures on the moon from 240° F. above zero to 270° F. below zero (510° degrees) is so great that men could not survive there. The daily range is much less here on earth. Perhaps the greatest variation on earth occurs in deserts where the daytime temperature may be 100–110° F. above zero, and the nighttime temperature 50–60° F. above zero, a range of some 60 degrees.

Before thermometers could be put on the moon, temperatures were measured with a thermocouple, a device that is extremely simple in principle. It consists of wires of two different materials that are fused together at their ends. One wire is made of bismuth, a pinkish-white, brittle metal. The other wire is made of an alloy of bismuth combined with a small amount of tin. The other ends of the wires (those that are not fused) are fastened to a sensitive instrument for measuring electricity. A thermocouple used for measuring moon temperatures is so small and weighs so little that a thousand of them would weigh only as much as a drop of water.

When energy falls upon the junction of the two wires, a small electric current is produced. This current causes a hand to move on a very sensitive galvanometer, an instrument that registers an electric current. The greater the motion of the hand, the higher the temperature.

In use, a thermocouple is placed at the focus of a telescope so that radiation (light, ultraviolet, and heat waves) from a part of the moon can be directed upon the junction of the two wires, the point where they are fused together. Such devices are extremely sensitive. When used with a 100-inch telescope, a thermocouple could register the heat of a single candle located 100 miles away, if there were no intervening atmosphere.

Thermocouples are not selective; that is, they register visible light waves, ultraviolet waves which are extremely short, and long infrared waves, which are often called radiant heat. In order to measure radiation from the moon itself, filters must be placed in front of the thermocouple. The filter, which is a thin sheet of glass, removes the sunlight reflected by the moon. In practice, two measures are made, one with a filter (sunlight removed), and the other without a filter (sunlight included). The filtered reading indicates the energy radiated by the moon's surface alone. When he has this knowledge, a scientist can compute the temperature of the moon.

You can make a thermocouple that will generate electricity. Obtain two 8-inch lengths of wire; one wire should be copper, and the other one iron. The iron wire used to hang pictures works very well. Twist the ends of the wires together tightly to make a loop. Cut the copper wire at the middle, and fasten the ends to a galvanometer.

Insert one of the junctions in a container of cracked ice sprinkled with a teaspoonful or so of salt, and hold the other junction in a match flame. The point on the galvanometer moves, indicating that an electric current is flowing in the circuit. You may be able to use a galvanometer in your school's science laboratory.

If you do not have a galvanometer, one can be made as follows. Wrap 50 turns of copper wire (number 18 bell wire works well) around a quart soda bottle. Slide the wire off, and bind the coil together with a short length of wire, or a piece of any kind of tape. Magnetize a razor blade by stroking it along a magnet. Erect the coil vertically on a block of wood. Suspend the magnetized blade from a thread inside the coil. When a current flows through the coil, the razor blade moves. The ends of the coil can be twisted directly to the piece of iron wire for the thermocouple.

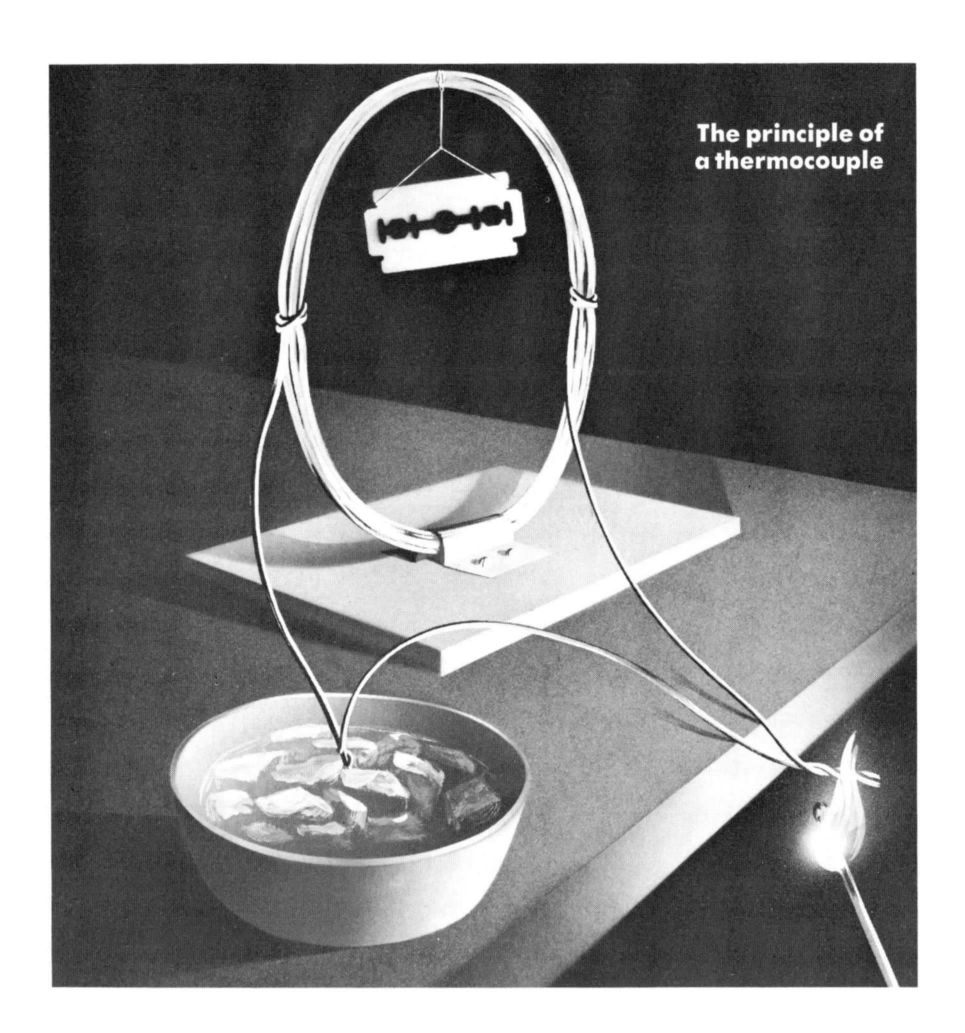
**The principle of a thermocouple**

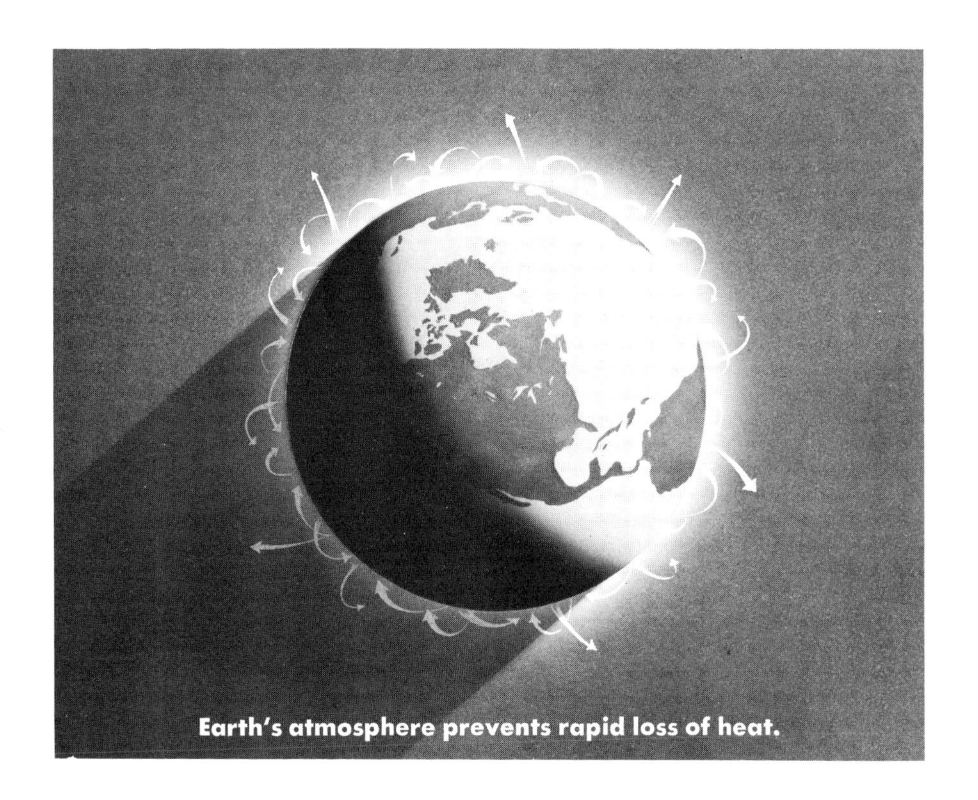
Earth's atmosphere prevents rapid loss of heat.

Earth is surrounded by an atmosphere that contains a large amount of water vapor. The water vapor permits radiant energy from the sun, which is short wave in nature, to pass through to earth. But the water vapor does not permit the passage of long-wave heat from earth. The water vapor acts as a blanket that prevents the loss of heat from earth. The moon has a very thin atmosphere, if any at all; and it has no water vapor. Therefore heat is lost from the moon very rapidly.

The rapid loss of heat from the moon is shown dramatically during a lunar eclipse. In the partial phase of an eclipse, the temperature drops from 240° F. to −148° F. This occurs in less than one hour. During a total eclipse, which lasts another hour, the temperature may drop another 18 degrees or so. Then, as the last partial phase

is reached, the temperature returns rapidly to 240° F. or thereabouts.

The temperature ranges that exist on the moon are extreme indeed. These extremes would be experienced continually by anyone moving across the lunar surface. Because of the absence of water vapor, and of any appreciable atmosphere, heat would not be distributed. Those places in direct sunlight would be unbearably hot; while adjacent places in darkness would be bitterly cold. The moon is certainly an unpleasant place.

# The Atmosphere of the Moon

It would not be correct to say that the moon has an atmosphere, in the sense that earth has an atmosphere. However, neither would it be correct to say that we positively know that there are no molecules of gases on or near the lunar surface. There is considerable possibility that the moon possesses some molecules of gases.

But astronomers are generally agreed that the moon has no atmosphere, in the strict sense. On the other hand, earth has a considerable atmosphere. It is surrounded by some 5,000,000,000,000,000 tons of gases, mostly nitrogen and oxygen, with relatively small amounts of helium, argon, neon, krypton, xenon, and carbon dioxide. If the moon has an atmosphere, then it must be very thin indeed. It could not be any more than 1/10,000 of earth's atmosphere, and very likely much less.

Countless observations support this conclusion. During one day the moon moves about 13 degrees across the sky, from west to east. (There are 5 degrees between the points of the Big Dipper; therefore the moon moves through about two and a half times that distance in a day.) In this movement, the moon occasionally passes in front of a star, blocking it from view. The star remains invisible for an hour, more or less. We say that the star has been occulted, and the event is an occultation. By watching occultations carefully with telescopes, it has been found that stars disappear instantly; one moment the star is bright and clear, the next moment it cannot be seen. The star does not fade from view, as it certainly would if the starlight were coming through a lunar atmosphere. Such fading from view would be especially apparent if the starlight passed through an upper atmosphere which would be thin, and then through lower layers of atmosphere that would be more dense.

The moon moves 13 degrees in 24 hours; therefore, it moves about ½ degree in an hour. The width of the moon is also about ½ degree. The diagram shows why stars may be occulted for varying periods of time. For example, the star labeled *A* will be occulted for a shorter period than the star labeled *B* because star *B* must cross a wider part of the lunar disk. This star will be occulted for about one hour.

Astronomers know that there can be very little, if any, atmosphere on the moon because stars are occulted just when they should be, and they reappear just when they should. Using mathematics, astronomers can compute the *exact* time when the moon will occupy a specific

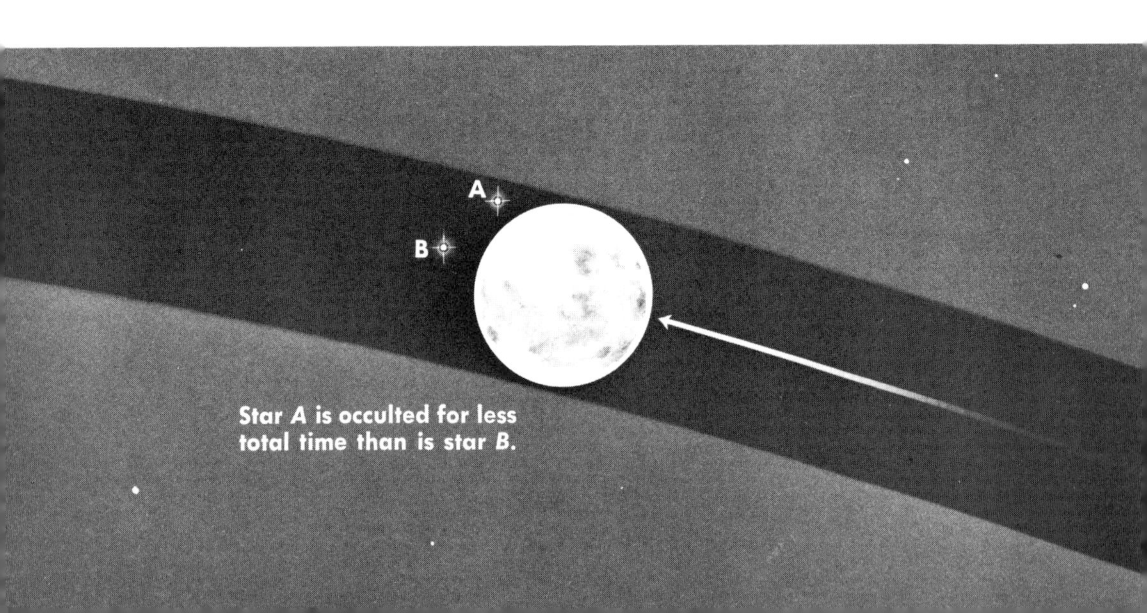


**Star *A* is occulted for less total time than is star *B*.**

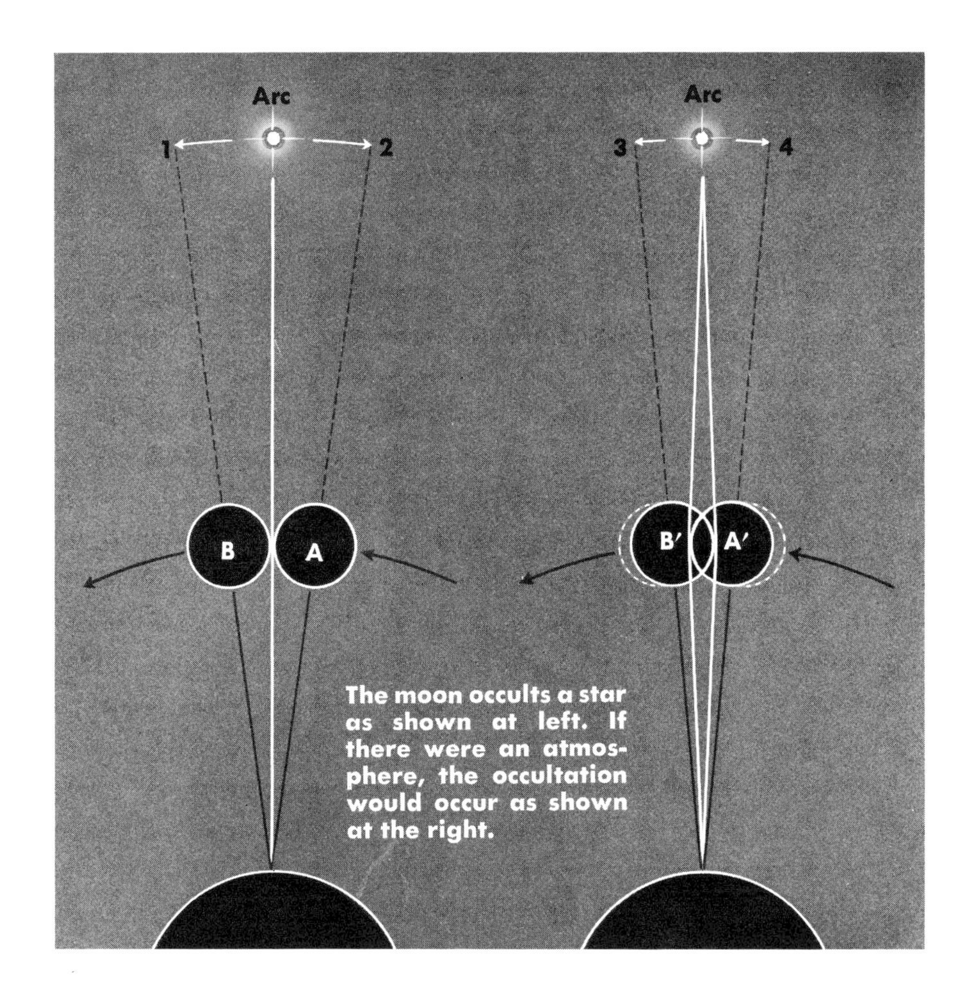


location in the sky and thus blot out a star. Also, using mathematics, astronomers can compute the precise instant when the star will reappear. And the star disappears and reappears at the exact time that it should. If there were an appreciable atmosphere on the moon, the star would be occulted late and it would reappear early. The diagram shows why. At the left we see what actually happens: the star is occulted when the moon is at A, and it reappears when the moon is at *B*. The observer would not see the star while the moon is moving through the arc 1-2.

The right side shows what would happen if the moon had an atmosphere. The atmosphere would bend (refract) light. Therefore, the star would be invisible only after the moon had moved to *A′*, even though the moon had moved beyond a straight line-of-sight to the star. The star would be visible again when the moon was at *B′*, before the moon had moved enough to clear a straight line-of-sight to the star. The star would be occulted only while the moon is moving along the arc 3-4.

No haze or clouds have ever been observed that obscure the entire surface of the moon. The lunar surface appears perfectly clear and sharp at all times. However, many careful and reliable observers have reported hazes that dimmed details of the lunar craters. But such observations were not confirmed because the sightings were very fleeting, and no photographic records or other reliable evidence to support them were obtained.

In 1956, Dinsmore Alter, an American astronomer, obtained photographs of a group of the three craters Ptolemaeus, Alphonsus, and Arzachel, in which he detected haze, especially in the Alphonsus crater. A Russian astronomer, Nikolai A. Kozyrev, studied the photographs, and he noted also that the details in the floor of the Alphonsus crater were especially blurred. Kozyrev decided to investigate further. In November, 1957, he obtained spectrograms (photographs that analyze light) which indicate that some kind of eruption occurred in the Alphonsus crater. The spectrograms, photographs made with a spectrograph (an instrument that separates light into the wavelengths that compose it), showed the momentary presence of materials such as carbon, substances that might have oozed out of fissures in the mountainous inner region of the crater.

The Russian astronomer suggested that it may be some time before observations of such activity on the lunar surface are again possible. However, the fact that such activity has been observed at all is very interesting. This is the first time since 1609, when the moon was first observed through a telescope, that gases on or near the lunar surface have been noted and the presence proved by such reliable evidence as spectrograms.

As soon as Kozyrev announced his discovery in various professional magazines and at international meetings of astronomers, astronomers in other parts of the world turned their telescopes toward the Alphonsus crater. However, after making photographs and spectrograms of the crater they were unable to note any permanent change in its appearance, and therefore it must be concluded that whatever gases may have been present when Kozyrev made his study have since disappeared.

There are other arguments strengthening the belief that gases of some kind exist in particular lunar locations. One of these arguments supposes that certain nonactive gases lie in the deep fissures. Probably lighter gases, those with atomic weights below 60, would have escaped. It is conceivable that heavier gases, once produced, may have been retained in the deep cracks. No gases were found in the lunar samples brought to earth by the Apollo missions.

Some of the elements in the lunar crust are radioactive and have been for a long time. When a material is radioactive, energy is released, and also new materials. Heavy rare gases such as krypton and xenon may have been produced by the splitting of uranium atoms. The gases, once produced, would not escape easily from the lunar surface because of their weight. Therefore, they may be quite abundant in certain areas of the moon.

There is also an argument for an atmosphere made of light gases, such as hydrogen. However, if there is such an atmosphere, its density would be only a few trillionths of the density of earth's atmosphere. It is likely that interplanetary space contains molecules of hydrogen gas, perhaps as many as one molecule per cubic centimeter. If so, then the moon must capture some of these molecules. But the moon cannot hold them, and they escape. It is probable that the moon loses as many molecules as it captures, and a balance is therefore maintained.

Observers have other arguments to support the belief that there is no atmosphere on the moon, or that the atmosphere is very thin. They point out that there is no twilight zone bordering the terminator. If the moon had an atmosphere, then there would be a gray region between daylight and darkness. There is none. On one side of the terminator there is bright, glaring light; on the other side there is deep, complete darkness. There is no twilight on the moon. As one moved from direct sunlight, he would enter full darkness, for there is no air to spread out, or diffuse, the light.

If the moon had an atmosphere, there would be a ring around the moon during every solar eclipse. A solar eclipse occurs when the moon blots out the sun. If there were an atmosphere about the moon —only 1/10,000 as much as earth's—the atmosphere would bend, or refract, sunlight so that observers would see a ring of light all around the moon. Indeed, when Venus, a planet with an atmosphere, moves in front of the sun, such a ring forms. It is a bright, jewel-like ring that encircles the black disk of the planet.

Since there is no appreciable atmosphere on the moon, there cannot be any appreciable amount of water. Water would evaporate, creating an atmosphere of water vapor, if nothing else. However, the vapor could not exist in any part of the moon out of direct sunlight, for the cold temperature would freeze it.

Why should the moon have an atmosphere? For that matter, why should earth have an atmosphere; where did it come from? There are many theories. One theory says that earth was very hot at one time, so hot that it was composed entirely of vapors. As earth cooled, the vapors condensed to become the rocks and seas. Other substances remained in the gaseous state. These became our atmosphere. This is a theory, just an idea, and we cannot say for sure whether such events occurred or not.

However, we do know that volcanoes here on earth bring certain gases to the surface, especially water vapor. Over a long period of time, millions of years, these gases of volcanic origin have contributed a great deal to our atmosphere.

Also, earth is bombarded by meteorites and it has been bombarded for millenniums. The meteorites become very hot on contact with earth, and gases are released. A large part of our atmosphere may have been obtained from the gases resulting from the heating of meteorites.

It is entirely reasonable to suppose that the moon has been exposed to such atmosphere-forming conditions. In other words, volcanoes and meteorites must have contributed to a lunar atmosphere. If this is so, and it appears to be, then what has happened to the moon's atmosphere? Astronauts have found no atmosphere on the moon, nor have instruments indicated any water, except for rare eruptions from below the surface. Also, the minerals found on the moon cause geologists to believe there never was water on the moon in any significant amounts.

All substances—solids, liquids, and gases—are made up of molecules. All molecules are moving, unless the temperature is absolute zero (−459.4° F.). The molecules in liquids move faster than those in solids, and the molecules in gases move faster than those in liquids. And, the hotter the gas, the faster the molecules move.

Compared to earth, the moon has a rather low escape velocity, 1.5 miles per second. Any object, such as a rocket or a molecule of a gas, can take off from the moon when it reaches this speed. This means that if a molecule on the moon moves 1.5 miles per second, it will reach escape velocity. If it moves away from the lunar surface, it can escape the moon's effective gravitational attraction. The molecule moves into space. At the temperature which sometimes prevails on the moon (240° F.) some molecules would reach 1.5 miles per second, and they would therefore escape. When the loss of molecules continued over a long period, the moon would have lost the gases it had acquired.

We can be sure that earth, with a higher escape velocity than the moon's, is losing molecules of its atmosphere for the same reason. Occasionally, molecules in our atmosphere reach earth's escape velocity, some 7 miles per second, and escape our gravitational field. These molecules probably move in orbits about the sun.

The moon probably retains some few molecules of gases. However, the number is so small that atmosphere on the moon is negligible.

# Mass and Density of the Moon

In the classic equation for universal gravitation, $F = \frac{G\,Mm}{r^2}$ Sir Isaac Newton illustrated the way the astronomer and the mathematician obtain knowledge that is otherwise quite obscure. Newton showed modern man the way in which he could determine information essential to successful flights of satellites, to studies of the moon, and to flights of probes designed to explore other worlds. (See "Why the Moon Stays in Orbit.")

When scientists know the mass, density, and gravitational force of a planet, they can then compute how fast a satellite or probe must go, the amount of fuel required, and other essential details.

The mass of a body gives us some idea how much matter the body contains. Density tells us how tightly the material is packed. For example, the density of lead is greater than the density of aluminum—matter is more concentrated in lead than it is in aluminum.

Gravitational attraction is related to mass. By studying earth's gravity, we have found the mass of earth to be $6.6 \times 10^{21}$ tons (this is 66 followed by 20 zeros, or 6,600,000,000,000,000,000,000 tons). In the metric system, the figure for the mass of earth is $5.98 \times 10^{27}$ grams. In astronomy, as in other sciences, the metric system of measuring is used rather than the English system. The reason is a simple one: the metric system is built on 10's, so that each unit is one-tenth that of the next higher unit. To multiply in the metric system, you simply move the decimal point to the right; to divide, you move the decimal point to the left. The English system is relatively awkward, for there is no similarity in the relationships of units; for example, there are 12 inches in a foot, 3 feet in a yard, 16½ feet in a rod, 5280 feet in a mile. An equally clumsy relationship exists in the tables of weight and volume.

When we know the mass of a body, we can then find its density. Density tells us how much mass there is in a certain volume. Usually the density is compared to water. Suppose we have a quart of water, weighing 2 pounds, and a quart of iron, weighing 15.6 pounds. The iron is 7.8 times heavier than the water; therefore, the density of the iron is 7.8.

To find the density of earth we need to determine its volume. In the metric system, the volume of earth is $1.083 \times 10^{27}$ cubic centimeters. Dividing the mass by the volume gives:

$$\text{density} = \frac{5.98 \times 10^{27}}{1.083 \times 10^{27}} = 5.5 \text{ (approx.)}$$

In other words, earth weighs about 5.5 times more than it would if earth were made entirely of water.

We can find the mass and density of the moon also. However, the way in which this is done is quite different.

Suppose that earth and the moon were connected by a metal rod that reached from the center of one body to the center of the other. And suppose that the mass of earth equaled the mass of the moon. Then, if the rod were supported at the middle, the two bodies would balance each other. The arithmetic works out like this:

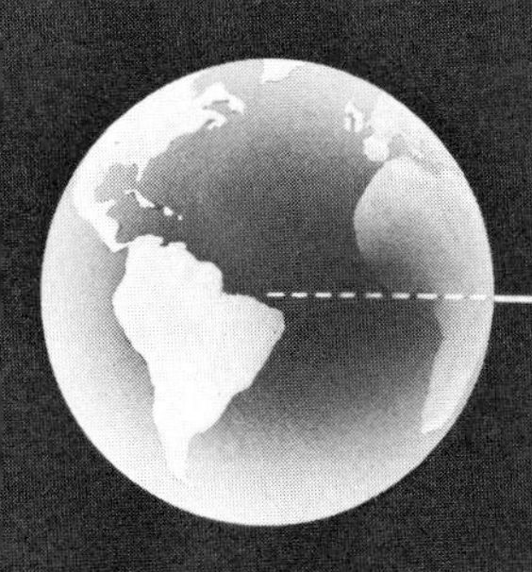


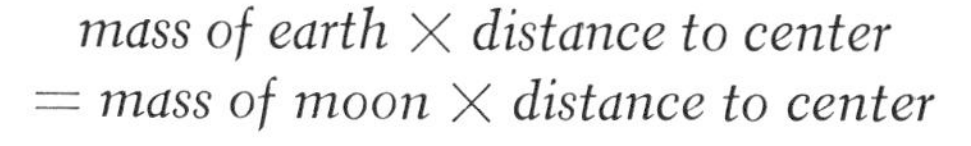

*mass of earth* × *distance to center*
= *mass of moon* × *distance to center*

Suppose that the mass of the moon were only one-half the mass of earth, then the balance (center of gravity) would be closer to earth. The mathematics would be the same:

*mass of earth* × *distance to balance point*
= *mass of moon* × *distance to balance point*

So far we know the mass of earth. If we can find the balance (center of gravity) of the earth-moon system, we can find the mass of the moon. This is true because, if we know the location of the center of gravity, we know the distance from earth's center to it, and we know the distance from the moon's center to it. Only one thing would be unknown—the mass of the moon. Let us see how astrono-

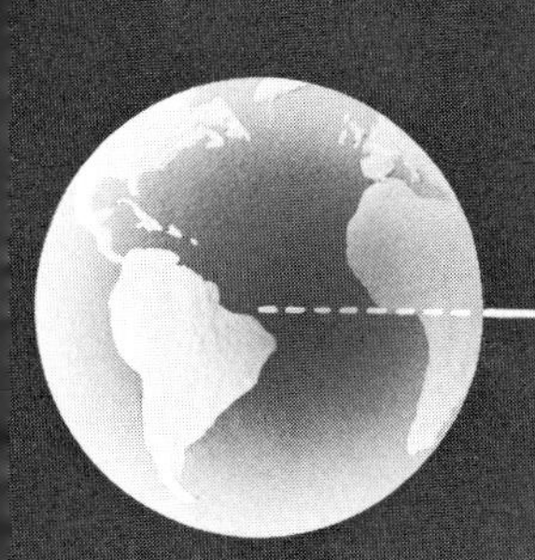

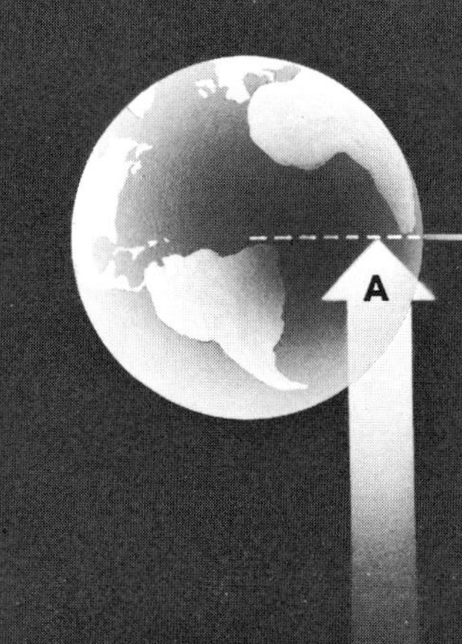


The "center of balance" of the earth-moon system is 2903 miles from earth's center.

mers determined where the center of gravity of the earth-moon system is located.

Strange as it may seem, we obtain the mass of the moon by observing the sun. We know that earth moves in an elliptical path around the sun, and when we say earth, we mean the center of earth moves along this imaginary line. But we know that it is the earth-moon system that moves around the sun and not earth alone. Therefore, it is the center of gravity of the earth-moon system that moves on the elliptical path around the sun. Because the mass of earth is much greater than the mass of the moon, the center of gravity must be very close to earth. Indeed, it is within the earth itself. Suppose we identify it roughly as A in the diagram.

To be accurate, the moon does *not* go around the center of earth. It goes around A, the center of gravity of the earth-moon system. At the same time, earth is going around the same point, though it moves in a much smaller orbit. Both bodies move about their common center of gravity, but the moon cannot make earth travel in as large an orbit as the much larger earth imposes on the moon.

As earth goes around the sun, the earth is sometimes ahead of the center of gravity; sometimes it is behind the center of gravity. The sun is seen against the sky at a point directly opposite the location of earth in its orbit. Therefore, the sun will sometimes appear to be moving a little fast or a little slow in its path through the sky. The variation is very slight, but sensitive instruments can measure it. Once the measurement is made, mathematicians can calculate the location of the center of gravity of the earth-moon system. It is 2903 miles from the center of earth. Now the mass of the moon can be computed.

This distance, 2903 miles, is 1/82 of the total distance between the center of the earth and the center of the moon; therefore the mass of the moon must be about 1/81 that of the earth. (The distance from earth's center to the center of gravity, 2903 miles, is one part of the 82; therefore the mass relationship is 1 to 81.)

We know the diameter of the moon is 2160 miles (its radius is one-half of this, or 1080 miles), so we can find the volume. In cubic miles, the volume of the moon is:

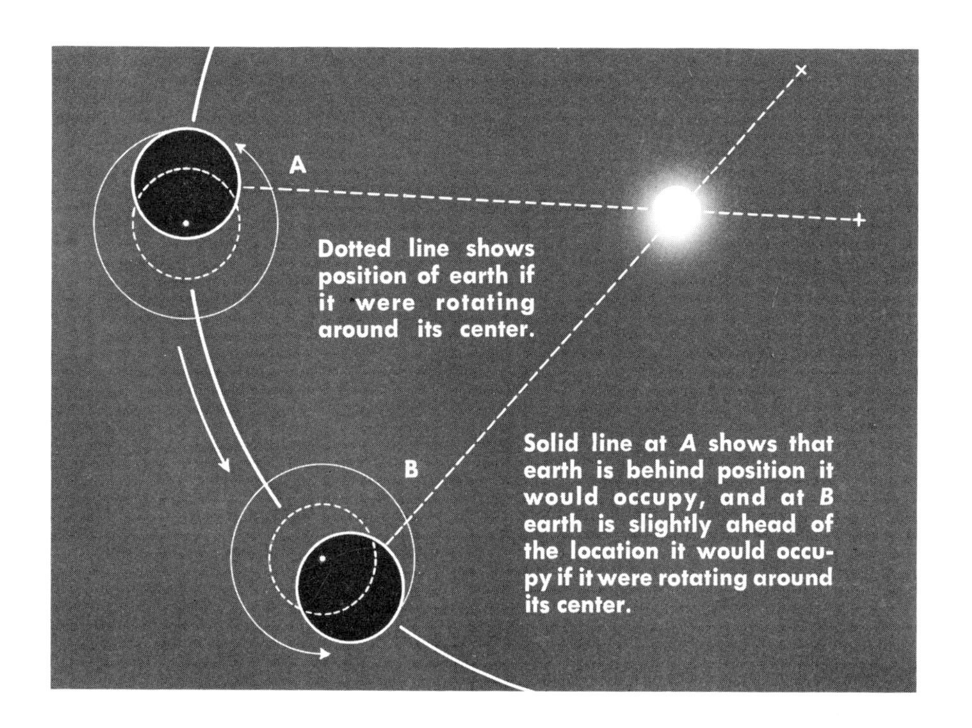

$$\text{volume of moon} = \frac{4}{3}\pi r^3$$

$$= \frac{4}{3} \times 3.1416 \times 1080^3$$

$$= 5{,}290{,}790{,}400 \text{ cubic miles}$$

This is about 1/50 the volume of earth. Now that we have the mass of the moon and its volume, we can compute its density. We find that the density of the moon is 0.6 times earth's density. You recall that the density of earth was 5.5; therefore, the density of the moon is 0.6 × 5.5, or 3.3 (3.3 times that of water).

The interior of earth is very dense, but the rocks that form a mantle or crust some 30 miles deep have a density nearly the same as that of the moon—about 3.5. This might mean that the moon is made of rocks similar to those in the mantle of earth. Early studies of moon rocks indicate this may be true. But there are concentrations of massive materials on the moon, peculiar formations that do not occur on the earth. They are called mascons, a coined word from the two words *mass* and *con*centration. Someone has said the moon may be somewhat like a raisin cake. The mascons would be the counterpart of the raisins.

Scientists do not know what causes these concentrations of massive material. They think molten rock from beneath mountainous areas of the moon flowed over a flat, lunar plain. The rock added to that area would produce a region of greater mass. Mascons were discovered because of the stronger gravitation exerted by these regions.

The gravitational attraction of a body depends upon the mass of the body. Suppose we give earth's gravity a value of 1. Then the gravitational attraction at the surface of the moon is 1/6 that at the surface of earth. If you weigh 120 pounds on earth, you would weigh only 20 pounds on the moon. Athletic records would be quickly shattered if a meet were held on the moon—pole vaults of 60 feet would be ordinary; a vault of 70 feet would be needed to place in the competition.

# Tides—Then and Now

If you live along a seacoast, you have probably kept records of the tides. You know that the ocean rises and then takes about six hours to fall. The intervals are more or less regular. If you have kept careful records, you know that the interval between two high tides is exactly 12 hours, 25½ minutes. This interval is exactly one-half the time required for the moon to appear successively on the same imaginary north-south line in the sky. This interval suggests that there is a close connection between the moon and the tides, and indeed there is. The height of the tides is related to the changes in the position of the moon. The highest high tides occur when the moon is new, when the sun and moon are on the same side of earth. We say the moon is in syzygy, from a Greek word meaning "joined together." These are the highest of the spring tides. Lower spring tides occur when the earth is between the sun and moon; they are in syzygy once more.

The lowest high tides occur when the sun, earth, and moon form a right angle. We say that the moon is in quadrature. These are

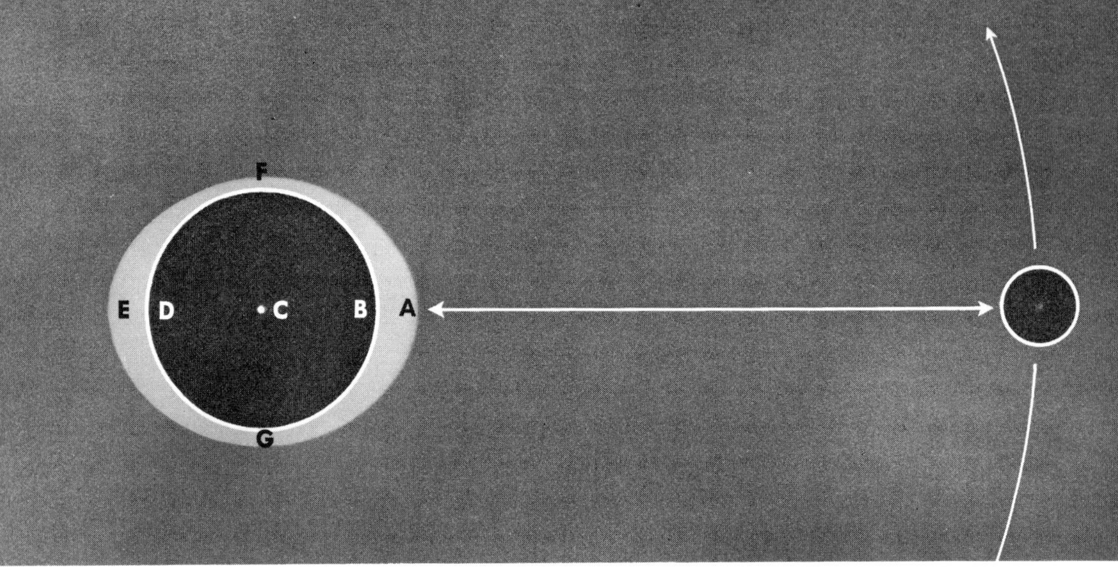


**Tides are caused by differences in the gravitational attraction of the moon on parts of the earth and its oceans.**

neap tides. Both neap tides and spring tides occur twice a month.

Another proof that tides are related to the moon is the fact that the highest tides occur when the moon is at perigee—when the moon is closest to earth. At this time, tides are about 20 per cent higher than they are when the moon is farther away.

The diagram explains the forces that cause the tides. Water on the moon-side of earth is shown at *A*, solid earth is shown at *B*, the center of earth at *C*, the more distant side of earth at *D*, and the ocean again at *E*. The moon attracts the ocean at *A* more strongly than it does the solid earth directly below it at *B*. Therefore, the water is piled up. The ocean at *E*, because it is so much farther away, is not pulled toward the moon as much as the solid earth; earth is pulled away from the water; and the water is therefore deeper during this part of the tidal cycle.

High tides occur at *A* and *E*. Low tides occur at locations removed 90 degrees. This would be at *F* and *G*.

The tide-raising force of the moon is very small indeed, compared to the force of gravity. The tide-raising force of the moon is about 1/9,000,000 that of earth's gravity, and the tide-raising force of the sun is only 1/19,000,000 that of earth's gravity, or only about ½ that of the moon's.

Suppose that earth were covered entirely with water and that earth always kept the same face toward the moon. Then the places of high and low tides would never change. High tide would be about two feet higher than low tide.

Now, suppose earth were turning slowly toward the east. Then the tides would move as a great wave toward the west. Really, earth would turn under the waters, and it would *appear* as though the waters were moving.

But earth is not covered with water. There are continents of all different shapes; there are islands, and bays, and all varieties of land formations. Therefore, tides are quite irregular. They may differ a great deal in the time of high water. Also, the height of the water varies greatly. In the Bay of Fundy in Nova Scotia, and also along the east coast of Patagonia, in South America, the tidal range is sometimes 50 feet. The high tide may come in as a great wall of water, called a tidal bore.

It has been suggested that the force of tidal waters should be harnessed and used to generate electricity. Incoming tides would turn generators. The water would be trapped by closing gates and then, at low tide, the water would flow through turbines back to the sea. Although the suggestion sounds like a good one, the cost of the installation would be much higher than the present cost of generating electricity by falling water and by steam power. The electricity would have to be stored to supply current during those intervals when the tidal flow was not of sufficient force to run the generators. To be really effective, generators must be able to supply a steady, uninterrupted flow of electricity. A century or so ago, some New England towns utilized tides to run their mills by placing power wheels directly in the tidal flow. Today there are no such projects in operation.

Sir George Darwin, son of the famous English scientist Charles

Darwin, made a detailed study of tides, and he calculated the past and the future of the earth-moon system.

Darwin said that at one time the earth and the moon were only about 10,000 miles apart. At that time the earth turned on its axis once in 5 hours. The moon went around the earth in a period of time just slightly longer than 5 hours. Because they were so near each other, each body must have produced tremendous tides on the other.

The great tides would have produced great friction; the waters would have dragged on the surface enough to slow down the rotation of earth. Therefore, earth would have lost energy. Energy cannot really be lost; whatever energy the earth-moon system possesses can only be transferred from one part of the system to another. Therefore, when earth "lost" energy of motion, the moon acquired it. The moon moved farther from earth, and the month, the time required for the moon to go around earth, became longer.

Darwin believed this process had continued for some four or five billion years. Indeed, we know that earth is slowing down even now. The day gets longer at a steady rate of 1/1000 second per century.

This process will continue; earth will slow down constantly, and the month will grow longer. After billions of years, earth will require 47 days to make a rotation, and the moon will require 47 days to go around earth. A balance will be attained. Friction of water against land, caused by lunar tides, will be zero.

But this balance will not last forever. Tides caused by the sun will slow down the rotation speed of earth and once more the day will be *longer* than the month. Now, as before, lunar tides will cause friction. But this time, because a day is *longer* than a month, the friction of lunar tides will accelerate earth, make it move faster. The moon will lose energy to earth, and so the moon will begin to move toward earth. It will get closer and closer.

But the moon will never collide with earth. As the moon nears earth, gigantic tides of lunar material will erupt, caused by earth's gravitational attraction. Various astronomers believe that these forces will be so great that the material of which the moon is made will be unable to hold together. The moon will shatter, they say, and the particles may form into a ring, or a series of rings, very similar to the rings of Saturn.

We advise you not to be unduly worried about this possibility, for the sun may cease to shine long before this dire event comes about. The shattering of the moon, if it occurs at all, will not happen for several billions of years.

# Reaching the Moon

It is unlikely that the moon will become a vacation resort, or that there will be regular flights to the moon now that men have landed there. However, we shall continue to explore this neighbor world of ours to find answers to the multitude of mysteries about it. Reaching the moon is not a new dream of men; there are indications that the Babylonians who lived five thousand years ago considered methods of getting away from earth. Apparently the Babylonians thought that going to the moon was a way of escaping one's enemies here on earth. Decorative art pieces have been found that show a man astride a great bird in flight toward a crescent moon; he is flying away from a man brandishing a great stick.

In the time of Alexander the Great a similar method for reaching the moon was suggested. The scheme was to harness two griffins, which were fabulous birdlike monsters that had the head and wings of an eagle and the body of a lion. Great chains were put around these creatures, and the chains were connected to a shallow basket in which the passenger stood. To steer the griffins, the passenger supported the carcass of a lamb on a long pole. The griffins were supposed to fly toward the lamb. Of course they never reached it. However, their continued efforts would carry the passenger to the moon.

After magnetism was discovered, and before people really understood this force, a method was suggested for using magnetism to reach the moon. A great ship was planned, one that was capable of carrying many passengers. A great magnet was to be fastened to the dome of the ship. Those who dreamed about such a conveyance believed that the magnet would continue to draw the ship forward as long as the magnet was kept above it.

Cyrano de Bergerac, a French writer who lived in the middle of the seventeenth century, suggested ways in which a man might reach the moon. For example, Cyrano noticed that dew rose in the morning from blades of grass and the leaves of shrubs and flowers. He suggested that anyone desiring to reach the moon should fill many cups with dew. The cups could be fastened about the waist of the moon-traveler. In the morning, the dew in the cups would rise, just as dew always rises. As the dew rose, the person would be carried to the moon, perhaps even to the sun. According to his writings, Cyrano made such a flight. He soared high above the earth. To keep from going too far, Cyrano broke a few of the cups, spilling the remaining dew. Although he started his flight in France, he landed in Canada because, as Cyrano said, earth turned beneath him while he was in flight.

Cyrano had another suggestion, one that used moving air. He suggested that a great box be built, the top and bottom of which should have openings. A great sail should be fastened above the upper opening. The upper half of one side of the box should also be open. This opening was to be turned toward the sunlight. The sunlight would heat air in the box. The hot air would rise out of the opening at the top of the box, strike the sail, and carry box and occupants aloft. Fresh, cool air would rush in through the opening at the bottom of the box. This was an early application of the jet principle.

Like other suggestions, Cyrano's convection-box idea assumed that there was air throughout space. The conception of a thin blanket of atmosphere surrounding earth was not brought forth until the latter part of the nineteenth century.

The great moon hoax of 1835, discussed in the following chapter, stimulated people throughout the world into thinking about the moon. Suggestions of ways to reach the moon came from far and near. For example, from Italy came a novel idea, one that utilized a graceful gondola. The sleek boat was fitted with a sail and with bellows. As long as the bellows were operated, a draft of air blew against the sail and kept the boat in motion. To be sure that the boat reached the moon, and no other destination, wheels were fitted to it. These wheels were cogged. The boat rested on two chains, and the cogs meshed into the links of the chain. The chain stretched from earth to the moon, so that there was no chance of the passenger's wandering aimlessly through space.

Perhaps the most famous way of reaching the moon was suggested in 1865 by Jules Verne, a French writer, in his book *From Earth to the Moon*. Verne suggested that the way to reach the moon was to be shot out of a cannon. Not just an ordinary cannon, but one 10 miles long. According to Verne's description of such an event, a great hole was dug in the earth, and the cannon was imbedded in it so that only the muzzle extended above the surface. The cannon was loaded with a great charge of gunpowder. A moon train was lowered into the barrel. When the gunpowder exploded, the moon train was shot into space through a great, concealing cloud of smoke and dust.

Jules Verne anticipated many events and conditions that have proved to be entirely correct. For example, two dogs were placed in Verne's moon train. While en route, one of the dogs, Satellite by name, died. A hatch was opened and the dog's body was thrown out. But, Verne tells us, the body did not fall. It traveled through space alongside the moon train. This is precisely what would happen if an object were thrown out of a fast-moving rocket.

Another man who anticipated modern events was Wan-Hoo, a Chinese merchant who lived some five hundred years ago. According to legend, Wan-Hoo carried out an elaborate experiment to test his method of reaching the moon. Wan-Hoo was strapped into a great chair. Forty-seven rockets were fastened to the chair back. Wan-Hoo held a kite in each hand to stabilize his flight. Then he directed his servants to ignite the rockets. They did, and according to the legend, that was the end of Wan-Hoo.

In 1957 men started the serious business of actually going to the moon when the first satellite was put into orbit around the earth. Four years later, on April 12, 1961, Yuri Gagarin made a trip around the earth in Vostok 1. He was the first man to go into an earth-circling orbit.

Experiments with satellites and space ships gave scientists the information they needed to send men to the moon. And that is precisely what they did in July 1969. At 10:56 P.M. E.D.T. on Sunday, July 20, Neil Armstrong, commander of Apollo 11, stepped to the surface of the moon from Eagle, the lunar ship that had landed shortly before.

In 1969 men walked on the moon. In the years ahead a lunar astronomical observatory will be built. There are many reasons why man wants to establish such an observatory on the moon. The usefulness of telescopes on earth is limited, for whenever an astronomer looks out to the distant stars, he must look through earth's atmosphere. The atmosphere always contains some water vapor and, equally troublesome, it always contains some dust particles. These interfere with the passage of light and tend, therefore, to obscure the objects under observation. Also, earth's atmosphere is never heated evenly throughout. Layers of air of different temperatures vary in density. Dense layers bend, or refract, light more than layers of lower density. Therefore, images seen through earth's atmosphere are often distorted. In addition to these more or less natural conditions, a condition very disturbing to astronomers is produced by man himself. Atmosphere catches the light of advertising signs, automobile headlights, and the general glow that is produced by towns and cities.

Observers located on the moon would have none of these conditions to contend with. From the moon the sky is always deep black, both night and day. Our daytime sky is bright only because particles in the atmosphere catch and reflect sunlight, diffusing the light and spreading it in all directions. But with no atmosphere on the moon, or very little, the sky is deep black at all times. To the astronomer, the moon has perfect seeing conditions, conditions free of the many distortions and distractions related to an atmosphere.

From a lunar observatory the sun would appear as a blue-white disk with rosy billows of gas projecting from the edge of it. Stars would be visible in all parts of the sky. Even close to the sun, the stars would glow with unblinking brilliance. And perhaps, even more important, the astronomer would receive *all* the light of the stars. On earth much of a star's light is filtered out as the light passes through the atmosphere. Indeed, much of the light in the violet and ultraviolet region never gets to earth at all. This lack makes it impossible for us to obtain full information about the stars.

The lunar observatory may have to be protected from bombardment by meteorites. On earth we are not concerned about this prob-

lem because the meteoroids (solid particles of matter in interplanetary space) that move toward our earth glow brightly and vaporize because of intense heating. The heating results from friction with our atmosphere. On the moon there is no atmosphere to heat and vaporize meteoroids, and therefore the particles may well bombard the surface continually. Some astronomers estimate that a million of them strike the moon in twenty-four hours. Admittedly, these would be very small. However, they might be an important obstacle to the successful establishment of an astronomical observatory on the moon.

As seen from a lunar observatory, earth would be a very imposing sight as we know from photographs made by lunar probes. It is a bluish globe with brilliant white polar caps, and is four times the size of a full moon—very impressive indeed.

From the nighttime side of the moon during the new-moon phase, earth is fully lighted. It reflects sunlight back to the moon which is fifty times brighter than moonlight on earth. When earth is between the moon and the sun, earth appears as a large, black disk four times the diameter of the sun. A bright silvery ring tinged with red surrounds the edge of the disk—sunlight refracted by earth's atmosphere.

From an observatory on the moon, the sun would appear practically motionless in the sky. It would rise in the east. But it would take about 7½ days to reach the meridian, the imaginary line in the sky that passes directly overhead. Then another 7½ days would be required for the sun to reach the western horizon.

Time in our lunar observatory would pose some problems. In observatories here on earth, time must be measured precisely. In order to do this, many observatories use a pendulum clock. The time needed for a pendulum to swing back and forth depends upon the length of the pendulum and the gravitational attraction of the place where it is located.

Here on earth a pendulum 29.1 inches long will require very nearly one second to make a swing. The length of the pendulum will vary a bit from place to place, for the force of gravity differs slightly from one location to another.

If this same pendulum timepiece were used on the moon, it would lose time rapidly. The gravitational attraction of the moon is only one-sixth that of earth, therefore the pendulum should be only one-sixth as long, about 5 inches.

However, it is probable that a different kind of timepiece will be used on the moon. A quartz crystal clock may be used. This keeps time by the steady vibration of quartz crystals. Or, perhaps, an atomic clock that uses the steady and regular release of particles from a small amount of a radioactive substance, such as cesium, will time the activities of a lunar observatory.

# Mythology and Superstitions

We know that the moon reflects sunlight to earth, and also that the gravitational attraction of the moon causes tides. We are sure of these two effects. The moon is often associated with other events that we are not so sure of—with harvests, rainfall, fishing, and even with birth and death. Scientists do not believe that the moon is related to these things; however, considerable folklore is built around such ideas.

Everyone knows about the man in the moon. Many people imagine that the light and dark regions trace out the eyes, nose, and mouth of a face. In oriental countries people imagine that they see the figure of a rabbit in the light and dark patches. In Scandinavian countries the light and dark areas of the moon are associated with the nursery rhyme that has come down to us as "Jack and Jill."

According to the legend, the moon stole two children from their parents and carried the children to heaven. The names of the children were Hjuki and Bil. The children had been drawing water from a well. They carried the bucket on a stick which they supported on their shoulders. The children, their bucket, and the pole were placed in heaven so that they could be observed from earth. These various parts of the story, the children and the buckets of water, refer to the dark regions (*maria*) of the moon. Even today, Swedish peasants say that the dark regions are the boy and girl carrying a bucket of water between them. From the story of Hjuki and Bil, we have the nursery rhyme:

*Jack and Jill went up the hill*
*To fetch a pail of water;*
*Jack fell down and broke his crown,*
*And Jill came tumbling after.*

Hjuki in the Scandinavian myth has become Jack, and Bil has become Jill.

The fall of Jack, followed by the fall of Jill, is related to the disappearance of lunar formations as the moon moves from full to last quarter and to crescent.

The carrying of water, and its spilling, probably have some relation to the widespread belief that there was a close connection between the moon and rainfall.

Many people of the world have believed that the moon was a god, or goddess, to be kept happy lest some great calamity should happen. Even today, there are tribes in Africa, Australia, and Indonesia that worship the moon. Much of the reverence accorded the moon springs from the lunar phases. Since the beginning of recorded history, and perhaps long before, men associated growth and happiness with the waxing, or growing, moon. In a similar fashion, the waning, or fading, moon has been associated with dwindling life; even with death itself.

Perhaps you know of people who even today regulate their lives by the moon. They believe that the time to plant seeds is when the moon is growing. In some South American countries the natives believe that a waxing moon gives energy to all living things, so mothers expose their newborn infants to direct moonlight.

In various parts of the world, people believe that trees should be cut when the moon is waning. They believe that the trees are slowing down their growth at that time, and that they are therefore easier to cut then.

There are many people who believe that the moon affects the growth of other plants, as well as trees. And there are some who are convinced that the moon affects the minds of people. Even as recently as the 1930's, the death of a young girl was blamed upon the moon. Newspapers reported that the girl was driven insane by the rays of a full moon.

A professor of an outstanding university reported that certain psychiatrists have long accepted the idea of "moon madness." He said that the moon is thought to have serious effects on people who have nervous disorders.

Such a connection is not believed to exist by most psychiatrists. However, it is possible that the moon does affect life in ways that are not clearly explainable.

Studies to determine the electrical rhythm of plants, as well as of human beings, have been made. The researchers report that there *may* be some connection between activity of life and the phases of the moon. This is an interesting possibility, and one that is being explored more thoroughly.

As mentioned earlier, the moon has had a prominent place in the religions of people throughout the world. For example, the American Indians worshiped the sun and moon as their main gods. They believed that the sun was the brother and moon was the sister of a family that lived in an ancient tepee. The brother and sister did not get along well together. They quarreled often. During an argument, the sister took a smoldering stick from the fire and ascended through the smoke hole of the tepee. The brother was angered that she escaped so easily. He grasped a larger, burning stick and pursued her. Even today, the girl moves across the sky. Later on, along comes the brother, still pursuing her. Occasionally, according to the legend, he catches her. Then there follows in the sky a great combat which results in an eclipse.

Lunar eclipses have been associated with a sleeping moon that must be aroused from slumber. Therefore the natives made as much noise as possible. For example, it was reported over two hundred years ago that the Cherokee Indians behaved very strangely during an eclipse. They ran wild like lunatics, firing guns, whooping and screaming, beating kettles, ringing bells, and making all possible noise. To them, eclipses were caused by a huge frog that gnawed away the edge of the moon. The moon would be destroyed completely if the frog were not frightened away.

These Indians, and people in other parts of the world, often connected the moon with weather changes. When the moon changes, they said, the weather changes. Actually, the moon is changing all the time—it waxes to full, and then wanes to new. There are a multitude of different lunar phases in addition to the new, first quarter, full, and last quarter phases referred to in almanacs. These phases occur about 7½ days apart; that is, it takes about 7½ days for the moon to change from new to first quarter, and so on. People who believe that weather changes are associated with moon changes do not usually keep records. If they did, they would find that the connection is questionable. Also, people who argue that the two are related are quite satisfied, even though the weather change may occur three or four days after the moon change.

There are many old sayings that relate weather to the moon. Some of these are given below:

*The full moon eats clouds.*

*Clear moon, frost soon.*

*Full moon in April brings frost.*

*Frost in the dark of the moon kills buds and blossoms;*
*frost in the light of the moon will not.*

*Pale moon doth rain.*
*Red moon doth blow.*
*White moon doth neither rain nor snow.*

*When the moon is visible in daytime, the days are cool.*

*When the sun runs low, expect cool or cold weather.*
*When the moon runs low, expect warm weather.*

*If on her cheeks you see the maiden's blush,*
*the ruddy moon foreshadows that winds will rush.*

*When the horns of the moon are sharp,*
*expect dry weather.*

*In the planting season no corn must be planted when a*
*halo is around the moon.*

*Sowe peason and beans in the wane of the moone,*
*Who soweth them sooner, he soweth too soon;*
*That they, with the planet, may rest and rise,*
*And flourish with bearing most plentiful wise.*

*Go plant the bean when the moon is light.*
*And you will find that this is right;*
*Plant the potatoes when the moon is dark,*
*And to this line you always hark;*
*But if you vary from this rule,*
*You will find you are a fool;*
*If you always follow this rule to the end,*
*You will always have money to spend.*

This is only a sampling of sayings that express the way people believe the moon affects weather. However, not everyone believes there is a connection between the two, for here is a saying that expresses a contrary viewpoint:

*The moon and the weather*
*May change together;*
*But change of the moon*
*Does not change the weather.*

*If we'd no moon at all,*
*And that may seem strange,*
*We still should have weather*
*That's subject to change.*

The first crescent moon seen in the western sky immediately after sunset is often associated with weather conditions. If the crescent is tilted upward so that it "cannot hold water," this implies that the entire month will be wet. If the crescent is almost horizontal to earth, it "can hold water," and the month will therefore be dry.

The crescent moon that occurs near the autumnal equinox is tilted so that it is nearly perpendicular to the horizon. This moon will not "hold water," and we would expect late September and October to be rainy always if the above belief were correct.

The crescent moon that occurs near the spring equinox is nearly horizontal to earth's surface and will "hold water." We would expect the latter part of March and April to be dry always if the belief were correct.

Weather is local; it may be quite different in a place removed by only a few miles. Therefore, how can lunar changes, which are the same for all parts of the world, affect such limited conditions? Probably there is no connection between the moon and the weather. This statement does not include beliefs about the "ring around the moon." We know that the ring results because of the bending of light by ice crystals in the upper atmosphere of earth. Therefore, rain is quite likely to occur soon after such a ring is seen. However, the ring is not lunar in nature. It is produced by a condition in our own atmosphere.

Not all mythology of the moon deals with beliefs concerning religion or weather. In fact, one of the most famous "lunar myths" of all times was about lunar creatures. Richard A. Locke, a newspaper reporter, wrote a fabulous but completely untrue account of lunar discoveries supposedly made by Sir John Herschel, a famous English astronomer. Locke reported, just as though it really were true, that Sir John set up a huge telescope 24 feet in diameter. The Palomar telescope, which is the largest in the world, is smaller than Locke's imaginary instrument. Locke reported that the 24-foot telescope was so powerful that Sir John saw fabulous creatures that looked somewhat like bats on the moon. Also, the surface was described in great detail.

These reports, which began modestly on the second page of the New York *Sun* on August 25, 1835, soon were printed on page one as the most important news story. This moon hoax, as it was proved to be later on, went into great detail concerning the many problems Sir John had to solve in order to design and build such a huge telescope. The lunarians were supposed to live in fabulous cities which could be seen easily through the instrument. The hoax caused great excitement throughout the world, for people believed that at long last living creatures had been discovered on another world.

The hoax was brought to a speedy end, however, when Locke closed the series by writing on August 31 that the huge telescope had been left facing east, and the sun burned a hole fifteen feet in circumference right through the reflecting chamber.

The truth about the moon is so exciting that made-up stories are not necessary to hold one's interest. For example, down through history the moon has played an essential role in shaping the lives of multitudes of people simply because it revolves around earth in regular intervals.

Because of its regularity, the moon has been used in timekeeping. Indeed, for centuries, people in various parts of the world have timed events by the moon. The American Indians made plans to hunt and harvest by so many "moons" ahead. They placed past events in sequence by the number of "moons" that had appeared since the events took place.

Today the moon is still used for measuring time in some parts of the world for certain purposes—usually religious. However, time is more frequently measured by the rotation of earth, and longer periods of time are measured by the revolution of earth around the sun. We do not try to fit the lunar months into the year. Neither do we begin the month with the phase of the moon, as people did long ago. Look at a modern calendar. In some months the first week contains a crescent moon. During other months the crescent moon may be in the second, third, or fourth week.

In days long past, such a method of timekeeping was quite impossible, for there were no calendars. The month was the interval of time between new moon or first crescent and the next new moon or first crescent.

The reckoning of time was often done by the priests of the church. The Hebrews assigned certain priests to go to a high tower and keep close watch for the first appearance of the thin crescent moon. The sighting made near the time of the beginning of spring was especially important, for this crescent moon ushered in the new year. The month that started then was called Nisan. It was the month in which Passover, the Festival of Freedom of the Jewish people, took place. Passover was, and is, a time of great rejoicing, and the news that the month of Nisan had begun was therefore spread far and wide among the Jewish people.

As soon as the priests saw the crescent moon, they blew a signal on a ram's horn. Runners below the tower, hearing the horn, set out for the neighboring hills. When they arrived at the hilltops, great fires were lighted; people living in the surrounding countryside watched the hills for the fires. When they saw them, there was great rejoicing. They knew that Nisan had begun, and they began preparing foods for Passover, the joyous celebration that would be held fourteen days thereafter.

Today the Jews still use the lunar, or moon, calendar for determining religious observances. Passover is the fourteenth day after the new moon in the cycle of phases during which a full moon occurs after the spring equinox. Therefore, Passover and Easter usually occur at about the same time. However, the year in the Jewish calendar is divided into lunar months, and the number of lunar months in the year does not come out even. To adjust the month and year relationship, an extra month is inserted into the religious calendar at regular intervals. During such times, the year has thirteen lunar months.

Similarly, the moon plays an important part in setting the date for Easter, which is a festive and joyous occasion in the Christian church. Originally, Easter was held on the third day after Passover. The Christian church teaches that Christ arose from the dead the third day

after the Last Supper. The Last Supper was the Passover feast. Passover can occur on any day of the week, therefore Easter could fall on any day of the week.

A large number of early Christians felt that Easter should always fall on a Sunday. They felt that Easter was a sacred occasion and should always fall on a sacred day. This dispute about the dating of Easter continued for several hundred years. Actually, it persisted until the fourth century, when a procedure was accepted called the Golden Number Rule. The basis for it was discovered by Meton, a Greek astronomer, around 433 B.C. Meton discovered that the phases of the moon recur after nineteen years on the same day of the month with a possible shift of one or two days, depending upon the number of leap years that have occurred. This nineteen-year series has become known as the Metonic Cycle.

Easter falls on the first Sunday after the first full moon after the vernal equinox (the beginning of spring). You can determine the date of Easter for any particular year by the following procedure: Divide the year by the number 19. Suppose that we use 1972 as an example:

```
     103
19/ 1972
    19
    --
      72
      57
      --
      15
```

The quotient, 103, is discarded. Add 1 to the remainder. In this case 1 is added to 15, giving 16. This is the Golden Number. Referring to the accompanying table of Golden Numbers, we find that Number 16 gives us March 30. This is the date for the first full moon after the beginning of spring in 1972. The thirtieth happens to be a Thursday, which means that Easter falls on April 2, the following Sunday.

| *Golden Number* | *Date of Full Moon* | *Golden Number* | *Date of Full Moon* |
|---|---|---|---|
| 1 | April 14 | 10 | April 5 |
| 2 | April 3 | 11 | March 25 |
| 3 | March 23 | 12 | April 13 |
| 4 | April 11 | 13 | April 2 |
| 5 | March 31 | 14 | March 22 |
| 6 | April 18 | 15 | April 10 |
| 7 | April 8 | 16 | March 30 |
| 8 | March 28 | 17 | April 17 |
| 9 | April 16 | 18 | April 7 |
| | | 19 | March 27 |

The earliest possible day for Easter is March 23. The only time Easter can occur on this date is during a year when the Golden Number is 14, and when March 22, the day of the full moon, falls on a Saturday. Easter must be the first Sunday *after* the first full moon. Easter and the full moon cannot both occur on the same day.

The latest possible day for Easter is April 25. This can occur only when the Golden Number is 6, and when April 18 is a Sunday.

These are only two examples of the many ways in which the moon has been used to date important events. Other religions throughout the world can, in a similar fashion, show that in the days of their struggles the moon and its phases were factors to be considered and used for purposes of dating.

Down through the centuries shepherds, herdsmen, cowboys, and ordinary city-dwellers have gazed at the moon, poets have been inspired by its beauty, and astronomers have charted its motions.

Except for Hermes, an asteroid, which at some time in the future may come within 220,000 miles of earth's surface, the moon is the

heavenly body closest to earth. Because it is so close, we know a great deal about our natural satellite, but we wish to know more. Man's curiosity to know more about the moon was partly satisfied in 1969 when he set foot on the surface, and when moon dust actually trickled through his gloved fingers. But man will not rest content until the many mysteries of the moon have been solved.

# Facts About the Moon

Diameter 2160 miles

Volume 1/49 that of earth

Mass 1/81 that of earth

Density 3.3 times that of water

Surface gravity 1/6 that of earth

Velocity of escape 1.5 miles per second

Velocity of motion 2287 miles per hour
3350 feet per second

Distance from earth 252,710 greatest
221,463 least
238,857 mean

Revolution period 29d. 12h. 44m. 2.8s. synodic (phase to phase)
27d. 7h. 43m. 11.5s. sidereal (true period)

Rotation period 27d. 7h. 43m. 11.5s.

Stellar magnitude –12.6 brightest

Albedo 0.07
(*reflecting power*)

# Further Readings


ALTER, DINSMORE. *Introduction to the Moon.* Griffith Observatory, Los Angeles, California, 1958.

ALTER, DINSMORE. *Pictorial Guide to the Moon.* Thomas Y. Crowell Company, New York, 1967.

ALTER, DINSMORE, and CLEMINSHAW, CLARENCE. *Pictorial Astronomy.* Thomas Y. Crowell Company, New York, 1956.

BALDWIN, RALPH B. *The Face of the Moon.* University of Chicago Press, Chicago, 1949.

BRANLEY, FRANKLYN M. *Pieces of Another World: The Story of Moon Rocks.* Thomas Y. Crowell Company, New York, 1972.

BRENNA, VIRGILIO. *The Moon.* Golden Press, New York, 1963.

COOPER, HENRY S. F., JR. *Moon Rocks.* The Dial Press, Inc., New York, 1970.

FISHER, CLYDE. *The Story of the Moon.* Doubleday and Company, New York, 1943.

GALLANT, ROY A. *Exploring the Moon.* Garden City Books, New York, 1955.

MASON, BRIAN, and NELSON, WILLIAM G. *The Lunar Rocks.* John Wiley & Sons, Inc., New York, 1970.

MOORE, PATRICK. *A Guide to the Moon.* W. W. Norton and Company, New York, 1953.

SLOTE, ALFRED. *The Moon in Fact and Fancy.* The World Publishing Company, New York, 1967.

WHIPPLE, FRED L. *Earth, Moon and Planets.* Grosset and Dunlap, New York, 1958.

WILKINS, H. P., and MOORE, PATRICK. *The Moon.* The Macmillan Company, New York, 1955.

Moon Set: 18 large photographs (8½ by 11¾ inches) made from Lick Observatory negatives. Key chart is included. Available at The Book Corner, American Museum-Hayden Planetarium, 81st Street and Central Park West, New York 10024, N. Y.

# Index

# About the Author

Dr. Franklyn M. Branley is well known as the author of many excellent science books for young people of all ages. He is also co-editor of the Let's-Read-and-Find-Out science books.

Dr. Branley is Astronomer and Chairman of the American Museum-Hayden Planetarium in New York City.

He holds degrees from New York University, Columbia University, and the State University of New York College at New Paltz. He lives with his family in Woodcliff Lake, New Jersey.

# About the Illustrator

Helmut K. Wimmer was born in Munich, Germany, and came to the United States in 1954. He immediately joined the Hayden Planetarium where he is a staff artist. Mr. Wimmer has illustrated many books on astronomy for young people.

He is also a sculptor and makes architectural models in his free time. He lives in New Jersey with his wife and children.